UNIVERSITIES ON FIRE

UNIVERSITIES ON FIRE

Higher Education in the Climate Crisis

BRYAN ALEXANDER

JOHNS HOPKINS UNIVERSITY PRESS | *Baltimore*

Published 2023
Printed in the United States of America on acid-free paper
9 8 7 6 5 4 3 2 1

Johns Hopkins University Press
2715 North Charles Street
Baltimore, Maryland 21218
www.press.jhu.edu

Library of Congress Cataloging-in-Publication Data

Names: Alexander, Bryan, 1967– author.
Title: Universities on fire : higher education in the climate crisis / Bryan Alexander.
Description: Baltimore, Maryland : Johns Hopkins University Press, 2023. | Includes index.
Identifiers: LCCN 2022029873 | ISBN 9781421446486 (Hardcover : acid-free paper) | ISBN 9781421446493 (eBook)
Subjects: LCSH: Environmental education. | Environmental sciences—Study and teaching. | Climatic changes. | Climatic changes—Study and teaching. | Communication in the environmental sciences. | BISAC: EDUCATION / Schools / Levels / Higher | ARCHITECTURE / Sustainability & Green Design
Classification: LCC GE70 .A44 2023 | DDC 363.738/74071—dc23/eng20221202
LC record available at https://lccn.loc.gov/2022029873

A catalog record for this book is available from the British Library.

Special discounts are available for bulk purchases of this book. For more information, please contact Special Sales at specialsales@jh.edu.

To the next generation of students, faculty, and staff, who must keep the academic flame alive amid the rising storm

To the current generation of academics: may we prove to be better ancestors

Our project is acknowledging that a future is coming when nature is no longer fully natural.

—Ruth Gates

We heard calls for community education and engagement in schools and across all sectors to increase understanding of climate science, the impacts of climate change, and what people can do. People wanted to see climate change education made compulsory.

—He Pou a Rangi
New Zealand Climate Change Commission

As a result of our inaction, we have three options: mitigation, adaptation, and suffering.

—Lonnie Thompson

CONTENTS

ACKNOWLEDGMENTS

This book only exists thanks to the help of a large swarm of people. Everything I write depends on them to various degrees. *Universities on Fire* is really a thick node where a series of networks intersect.

For terrific conversations, sharing of stories, reality checks, support, and inspiration: Doug Belshaw, Todd Bryant, Bernard Bull, Adam Bush, Tim Carroll, Zach Chandler, Sayeed Choudhury, Heather Christenson, Mary Churchill, Donald Clark, Elena Clark, Maree Conway, Shermon Cruz, Tanja de Bie, Bonnie Dede, Teddy Diggs, Stephen Downes, Steve Foerster, Vivian Forssman, Clyde Lee Graham, Jody Greene, Steven Greenlaw, Laurel Halbany, John Harney, Andy Havens, Tom Haymes, Joe Henderson, Paul Henley, Danial Jameel, Steven Kaye, Blair Kettle, Joshua Kim, Mark Kozitska, Neal Leary, Cliff Lynch, David Marble, Chris Mayer, Robert McGuire, Bill McKibben, Liz McMillan, Sally Mudiamu, Rod "The Militant Optimist" Murray, John O'Brien, Marius Oosthuizen, John Pettus, Mo Petzel, Stephanie Pfirman, Ruben Puentedura, Howard Rheingold, Roxann Riskin, David Rosowsky, Peter Rothman, Mike Roy, Mark Rush, Richard Sebastian, Heather Short, Meryl Shriver-Rice, George Station, Michael K. Thomas, Alex Usher, Vanessa Vaile, Ed Webb, and Andrew Zubiri. There are more of you too.

To the Last Humans Project for its vision and comradeship: Trent Batson, Randy Bass, and Phil Long.

Many people have been generous on social media, from my blog to Twitter, LinkedIn, Facebook, and Mastodon: Karl Aho, Ann Anderson, Maree Conway, Faolan C-P, Tanja de Bie, Nick Doty, Bart Édes, eminishteacher, Esty, floatingtim, Vivian Forssman, Travis Holland, Jens

Larsen, lmockford, Tara Magdalinski, Jonathon Richter, Sez, Jeremy Stanton, and Alex Usher.

Georgetown colleagues and students over the past several years have been a great help: Nafisa Isa, Eddie Maloney, Nikki Marks, Wesson Radomsky, and Andrew Zubiri. Students in my future of higher education seminars were terrific discussants. My thanks to the splendid Center for New Designs in Learning and Scholarship (CNDLS) and the Learning, Design, and Technology (LDT) programs for their support, especially letting me design and teach seminars there. The Big Rethink project helped ground my thinking about environmental justice in higher education. And students elsewhere helped, like those in Christine Dang's New York University media studies seminar or the inspiring leaders at SUNY Oneonta.

Readers in my online book club contributed to my understanding and analysis as reading after reading engaged the climate crisis: Steve Foerster, Joshua Kim, Chris Mayer, Sally Mudiamu, and more.

Members of the Future Trends Forum community have been a fine group to think with over our six years together. So many of you have helped make this book possible. I can single out Ofelia Mangen and Mathieu Plourde, yet there are many more. Thanks to Steven Gottleib for his energetic support throughout.

People playing games with me have been helpful, and I hope to see more books offer this line of acknowledgment in the future. For example, I appreciated a 2020 matrix game about geopolitics and climate change, organized by Stephen Aguilar-Millan. Its umpires and fellow players were delightful and helped nudge my thinking forward. Georgetown and George Washington University students contributed greatly to developing my future of the university game, which turned toward climate.

My supporters on Patreon have been terrific. There have been so many thoughtful and practical conversations with friends like Brett Boessen, Paul Henley, Joe Murphy, Joanna Richardson, and Vanessa Vaile.

Other organizations and institutions: Carbon Brief is a superb resource for clear explanations and extensive research. My colleagues at the Association of Professional Futurists have provided many fine conversations and resources.

I and everyone in the climate crisis movement owe a huge debt to great thinkers, visionaries, and selfless contributors to the cause. I'm thinking of the amazing writers Bill McKibben and Kim Stanley Robinson, who have showed us all ways forward with their words, and also put up with my barrage of questions.

Early readers, including Trent Batson, Mark Kozitska, and Mike Roy, have been both supportive and keen eyed in improving this book.

Greg Britton at Johns Hopkins University Press has been a grand editor and networker extraordinaire.

My family has been vital to this book's creation. Each brought their professional capacities to bear in helping sharpen my thoughts and expand my research: Ceredwyn, the contact tracer and emergency services guru; Gwynneth, the disaster planner; Owain, the historian. All are writers. All have also been supremely generous with time and emotional support. All put up consistently with the bizarre emissions from my brain. You all have all of my love.

ABBREVIATIONS

AI	artificial intelligence
AIACC	Assessments of Impacts and Adaptations to Climate Change
AMOC	Atlantic Meridional Overturning Circulation
AR	augmented reality
CIC	Council of Independent Colleges
CO_2	carbon dioxide
COP	United Nations Conference of the Parties
DAC	direct air capture of CO_2
GDP	gross domestic product
GHG	greenhouse gas (emissions)
GIS	geographic information system
IEA	International Energy Agency
IPCC	Intergovernmental Panel on Climate Change
ISP	internet service provider
LOCKSS	Lots of Copies Keep Stuff Safe
MLA	Modern Language Association
OA	open access (for scholarly publication)
OER	open educational resources
SDGs	Sustainable Development Goals (United Nations)
SRM	solar radiation management (a form of geoengineering)
SSP	shared socioeconomic pathway
VR	virtual reality
VUCA	volatile, uncertain, chaotic, and ambiguous
WEIRD	Western, educated, industrialized, rich, and democratic
WWA	World Weather Attribution

UNIVERSITIES ON FIRE

INTRODUCTION

Academia Wades into the Anthropocene

It was getting hotter.

—KIM STANLEY ROBINSON, *THE MINISTRY FOR THE FUTURE* (2020), OPENING LINE

Imagine a university under attack.

Nature besieges the campus. Floods from rising waters surge through the grounds, pounding buildings, silting up lawns, and crashing against libraries. Or fires roar through local forests and underbrush, filling the air in classrooms and fields with particulate-dense smoke, threatening to consume academic buildings. Or on another campus, a wet, sopping heat bursts the bounds of human tolerance. Students and faculty collapse from heatstroke, and the university is forced to air-condition the very outdoors. Or the grounds and the neighborhood dry out in harsh aridity; dust and sand sift onto the grounds and into buildings, while the institution hoards potable water.

Or consider this: perversely but unsurprisingly, humanity joins the climate in attacking colleges and universities. Imagine a local or national government mandating strong climate mitigation laws. In response, some in the academic community resist. Or the reverse occurs as a passionately decarbonizing university collides with a crisis-denying regional government. Anti-immigration activists attack a college for teaching or hosting climate refugees. Climate-driven economic disaster wracks a city, making financially durable urban universities targets of resentment, envy, or violence. Activists on campus target endowments, professors, staff, or fellow students as guilty of crimes against humanity and the earth for promoting carbon-emitting industries and practices.

In the face of such threats, and a great deal more, what is the fate of academia in that dangerous world?

To seriously answer this question, we must think through the possibilities—now—in the spirits of foresight and precaution. Then we can pose two questions. **First, how could climate change impact higher education, from individual students, staff, and faculty to institutions around the globe? Second, what does academia have to offer the rest of the world as civilization grapples with the developing climate crisis?**

The first question considers colleges and universities as something like passive objects in the era of global warming. Nature hits them. Campuses are not doers, but are done to. Responding and planning for all of that requires foresight and research. The second question posits the reverse because academics can also be active actors in this vast drama. Members of the higher education community can conduct research on the climate crisis, which can be of vital importance to the world, from studying ice cores and refining models that forecast oceanic transformations to analyzing how the human mind and human communities may react to climate stresses. Colleges and universities also teach, of course, meaning they play a crucial role in shaping the minds of generations. Additionally, these institutions play roles in their local communities, potentially helping and productively collaborating as the climate century advances. Beyond that, academics can be involved on the world stage, attempting to contribute to our civilization's overall response to the greatest challenge of our time.

Given that two-way relationship, in which academia contributes and is also subject to the crisis, we can start answering these two questions by exploring possible futures. Let us start by recalling higher education's footprint, recognizing the sheer magnitude of academia's material dimensions in the world. This can help us refine our sense of scale and get a better handle on potential threats. Globally, there are roughly thirty thousand postsecondary institutions at present. Their population includes around 220 million students and perhaps more than 6 million faculty members. Combined, that's larger than the population of many countries. Economically, estimates of how much money colleges and universities have range from $2 to $3 trillion. These figures sketch a sector with a considerable footprint on the world stage. Academia as a whole might count globally as a medium-sized nation or a single major corporation.[1]

Given this top-level sketch of academia as a human enterprise, imagine the following brief scenarios by which the climate crisis acts on it, and higher education's response:

- Students arrive at a campus where plants bedeck every building. Forests and gardens stretch between buildings. Electric, autonomous vehicles quietly course over a small number of solar panel–covered roads. A blimp soars overhead, bringing the day's mail. Large carbon-capture machines hum every hundred yards. Tall solar panel stations line a brook as it flows toward the giant, quietly swooping wind turbines on the other side of the grounds.
- Fires strike a college more and more often, consuming vegetation and buildings, causing lung damage to students and staff. Surrounded by flames in multiple seasons over years, the campus's buildings and public spaces are redesigned to reduce damages. A shield of open and empty ground surrounds the academic grounds, not offering any way for the next fire to carry itself into the college interior.
- One university is drenched in technology. Large screens dot the sides of buildings. Holograms appear on the grounds. Students, faculty, and staff wear glasses that connect them with the immersive internet. Robots are as present as humans. The goal is to make the off-campus world as rich and present as possible in order to reduce carbon-emitting travel. Yet in the next city over is a low-tech

college, where in-person interaction is the focus and the internet is blocked from campus. The reason here is to not rely on carbon-spewing servers and device factories, as well as to resist dependency on giant tech firms invested in aggressive economic growth: two opposing visions.

- A nation crumbles, ruined by multiple natural disasters in turn exacerbated by catastrophic human decisions. Its government declares a state of emergency. A decimated population struggles with survival, and many emigrate, joining the large and growing wave of international climate exiles. Colleges and universities, losing students, staff, and faculty, shrink, then close. Higher education here ceases to function institutionally.
- Colleges and universities set up climate change majors, centers, and programs—or defund them. Entire graduate schools and undergraduate colleges specializing in the climate crisis appear, while governments ban other institutions from researching or teaching the topic.
- University students condemn several professors for their alleged advocacy of carbon-intensive economics and research, occupying their offices and calling on the faculty members to resign or be terminated and their positions ended. Some staff and other campus faculty oppose this effort, while others join the protest, which soon gets national, then global attention. At the same time, the provincial legislature fiercely debates whether that same university's research into geoengineering should be supported and celebrated or if that academic program should be shut down and punished. Threats of violence occur, followed by the reality.
- Some professors cease long-haul and international travel for professional work after on- and off-campus protests. Instead, they share videos for their papers, posters, and presentations; hold discussions through telepresence robots, videoconferencing, multiplayer computer games, or immersive virtual reality; and host local in-person gatherings to which others travel by rail. Other faculty and staff oppose this movement and refuse to do so

themselves, charging the flight-shaming (Swedish: *flygskam*) protests with anti-intellectualism.

- Rising sea levels infiltrate water sources that previously held freshwater. Over time, trees, lawns, and gardens die as salt infests their groundwater. Buildings gradually subside into drenched soil. The institution's leaders quietly plan on migrating the entire campus to a new, inland location over the next decade. Sixty miles away, the same ocean's waters roll over an abandoned campus. That institution migrated from the site years ago. Nobody has purchased the grounds since.
- Scientific research stations are destroyed by extreme weather events, advancing deserts, and rising sea levels. Research into glaciers is stymied by those ice features collapsing or melting. Archaeologists lose access to irreplaceable sites. Academic teams carefully and extensively digitally scan natural sites for posterity before they change radically. Scholars conduct some research entirely by diving into virtual reality representations of pre-Anthropocene nature.
- Growing numbers of refugees from climate disasters and climate-caused political chaos arrive near universities. Some academics, politicians, and civil society leaders request that their institution host them on-site for humanitarian reasons, while others ask that campuses stay out of political issues beyond their purview and that they cannot afford to participate in for budgetary reasons. Without institutional permission, students throw open their residence hall rooms to refugees, and professors create online classes to teach migrants.
- As the twenty-first century proceeds, some universities and colleges close as their local or national economies are hit by climate stresses. Others merge on their own, or are compelled to do so, to consolidate precious resources. Some institutions relocate their campuses away from rising water or expanding deserts. Certain nations refer to this as an academic migration, while others call it a university dieback.

While some of these examples might sound hypothetical, traces of several have already occurred. Those that haven't yet are simple extrapolations from the present day, based on conservative estimates of forthcoming climate change.[2] Beyond these scenarios are many other possibilities, which this book will explore.

Returning to our two questions, we can start to answer them in another way. We can think of the whole climate crisis at a planetary scale, then fit academic institutions within that broader picture. Such a contextualization can allow us to link together multiple domains and sources of information, giving us insights across the familiar barriers separating an ivory tower from its neighboring world. At such a large scale, we can break down potential impacts by what speed they hit and through which causal chains.

Primary climate crisis impacts on academic institutions involve direct attacks on campuses, as noted previously. Sea level rise gradually gnaws at seafront properties. Serious storms, hurricanes, and tsunamis that increase in intensity, frequency, or both hit campuses on or near oceans. The storm surges that attend those events similarly attack academic built environments. Campuses adjacent to or near deserts, as well as those located in drylands, face a mix of threats, including local water sources drying out, rising amounts of dust and sand in the air and on grounds, and increasing frequency of dust and sandstorms, all of which threaten human health and the built environment. Increasing temperatures make it increasingly difficult or dangerous for humans to be outside of buildings, especially at institutions in warmer climates, notably tropical areas. In some areas, especially those in a widening band centered on the equator, rising temperatures make human existence outdoors challenging, then dangerous, especially when accompanied by humidity. Unusually high heat can impair human cognition and decision-making. Warming conditions can also afflict infrastructure near a campus, as when melting permafrost destabilizes roads, pipelines, and electrical power grids.[3]

Secondary impacts are those caused by the first-order ones, and that also hit academic institutions. As the climate changes, generally warming, it transforms ecosystems around the world, causing ripple effects

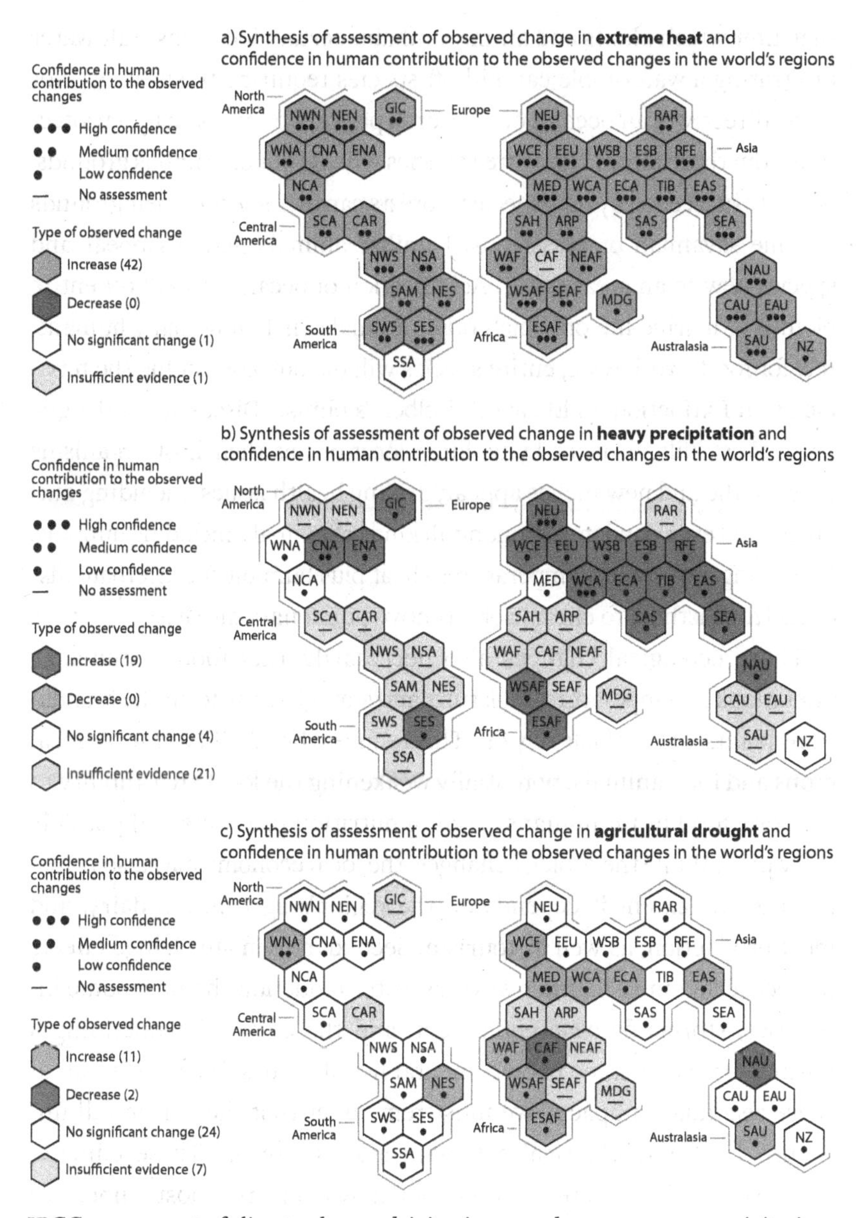

IPCC assessment of climate change driving increased temperature, precipitation, and drought worldwide. Figure SPM.3 from IPCC, "Summary for Policymakers," in *Climate Change 2021: The Physical Science Basis. Contribution of Working Group I to the Sixth Assessment Report of the Intergovernmental Panel on Climate Change*, ed. V. Masson-Delmotte et al. (Cambridge University Press)

that alter which plants and animals inhabit certain locations. Salt water infiltrating a water table can kill off species requiring freshwater.[4] Not being directly in an ocean's or a desert's path is not necessarily a protection from these forces. Some tree species will die out on campus grounds, possibly succeeded by new species. Forests can dwindle and their grounds become savannah or grasslands. Familiar animals may disappear and species new to an area appear. Acidification of oceans can kill off entire biomes of marine life on which other animals, including many humans, rely for food. Worldwide, entire species will die out, continuing the terrible Sixth Extinction, in Elizabeth Kolbert's phrase. Diseases like dengue fever, West Nile virus, or Lyme jump locations as their host organisms move or die and new niches appear, driving health issues, including outbreaks and even pandemics, among all kinds of animals, including humans. Increasing "wet-bulb" temperatures (heat plus humidity) exacerbate diseases from asthma to chronic obstructive pulmonary disease.

Those ecological changes also necessarily alter food systems, as alterations to temperature, moisture, and sunlight transform the foundations plants and animals rely on. This reduces the viability of some local crops and food animals, potentially weakening the local area's ability to generate food fit for humans, driving nutrition problems and possibly starvation. While the ecology changes, the local economy may also transform to the extent it depends on the region's agriculture, dairy, and meat production, as well as tourism. Secondary climate change effects hit local economies in other ways, as water reductions hamper some industries, storms damage equipment, and the costs of maintaining a business rise. For some people, mental health issues follow. With all of these secondary impacts, we must remember that they often fall unevenly. In the World Health Organization's words: "These climate-sensitive health risks are disproportionately felt by the most vulnerable and disadvantaged, including women, children, ethnic minorities, poor communities, migrants or displaced persons, older populations, and those with underlying health conditions."[5]

How humanity responds to these primary and secondary forces yields a tertiary level of climate change impacts, no less powerful for being so reactive. All of these second-order pressures, especially threats to food and water, can drive people to leave threatened areas, causing population flows that may amount to historic migration levels. Natural forces injure economies in many ways, from destroying property and increasing operating costs to disrupting supply chains. People living in poverty are especially susceptible to having to flee their homes, absent massive socioeconomic transformations. Competition for dwindling resources may drive increases in crime and suicide, as well as political unrest and violence, which is why many militaries now consider climate change to impact their operations while multiplying the force of others.[6]

These two forces of instability and migration can interact to produce more of each, as we know from both recent and general history that migrating populations often elicit political instability, at times building up to popular violence or outright war. Political decisions about climate change can significantly transform societies, as intended and through unintended consequences. Decisions to mitigate the crisis include support for installing or expanding alternative energy resources (solar, wind, hydro, and geothermal, as well as nuclear, which occupies a separate category); renovating power transmission and storage systems; encouraging or discouraging dietary practices and food systems; advancing or shrinking various transportation forms (air, rail, automobile, bus; petroleum fueled versus electrically powered); supporting or retarding climate science; building or resisting construction of sea barriers; moving or renovating government buildings, including at least one national capital so far; and encouraging or discouraging some forms of architectural design (carbon-neutral or carbon-negative buildings).

Private sector actions in response to the climate crisis are similarly diverse and occur on many of the same grounds. Decreasing prices for wind and especially solar power make those alternative sources more accessible; individual choices and changing social norms can drive their further adoption. Social norms and cultural expectations in general may change across many domains, from habits of eating to religious attitudes about humanity and nature, from consumerism to travel. At the same time, public and private decisions to avoid climate mitigation, or to continue the human practices that drove the crisis in the first place, can easily continue in various cultural forms, from resignation to climate denial to gleeful hostility. All kinds of continuities are already at work: extracting and burning coal and oil, fracking, driving internal combustion vehicles, flying jets, living in climatologically endangered areas, consuming products composed of or closely depending on greenhouse gas emissions, from food to clothing, shelter, and entertainment. Governments, businesses, and nonprofits support these habits to varying degrees and in different ways. Overall, we can consider climate change's tertiary effects as nothing less than a global struggle to redefine much of human civilization.

In that context, colleges and universities are now and for decades to come will be subject to all three levels of these climate crisis impacts to different intensities and at different times, depending on their physical location (primary level), how the immediate area around a campus and its population evolves (secondary), and how the culture and polity that they inhabit reacts and plans (tertiary).

It is worth noting that this threefold model can backfire due to the cumulative size of the challenges it describes. Why should we discuss academia at all when the entire crisis is so much larger, impacts so many more people, and is just that much more terrifying in its full planetary and existential scope? Some in postsecondary education would pose the

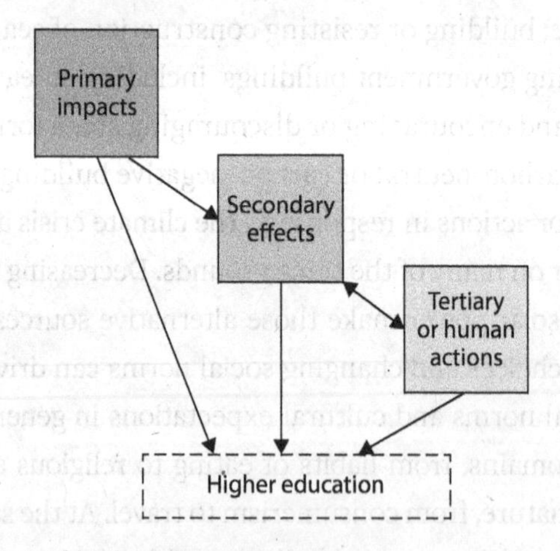

The climate crisis and higher education interact

same question in various forms. Higher education is not generally a home to large numbers of climate change deniers, but many academics have not been focusing on the climate crisis. Partly this is due to the nature of universities, which, according to tradition, offer a respite from the world to researchers and students engaged in learning and discovery. Those within academia can focus on bioinformatics or the Korean language and not have to devote much mental energy to external threats, including the Anthropocene, much as faculty and staff do not have to concern themselves on a daily basis with the operations of campus infrastructure.

This ivory tower explanation is to a degree a myth, as generations of scholars have sought to bring the world and its academy together, especially through community engagement and public intellectual practice. Moreover, on the student side of the academic community, Sara Goldrick-Rab and others have shown that good numbers of students cannot set aside the world's realities while studying. Yet this ivory tower belief persists nonetheless, both within and outside of university boundaries. The present volume is an attempt not only to cross those boundaries, not only to demonstrate that climate change will easily eradicate any such division between world and academia, but to show that the climate crisis has already breached those barriers. Its invasion will only widen in the coming decades. Not realizing this and not reacting strategically is ultimately self-defeating.[7]

Another reason to not think about the climate crisis and its impact on academia is that recent years have presented enormous stresses beyond climate change that have, understandably, soaked up our attention. The 2008 financial crisis hammered the economic fundamentals of many institutions worldwide, and a good number are still recovering financially. COVID-19 hit the world starting in 2019, reached pandemic status in early 2020, roared through societies throughout 2021 and the following years, and did not spare colleges and universities. Academia suffered on multiple levels, starting with infections, injuries, deaths, and mental trauma in the academic population, and moving on to dealing with economic, cultural, and political fallout. At the same time, multiple nations have had to deal with local crises that struck higher education, from the United Kingdom's Brexit process and Myanmar's coup to government repression in China's Xinjiang autonomous region. Rising authoritarianism and a crisis over what constitutes truth both present threats to academics and our work. While these stresses occur, and will doubtless be accompanied by new ones, the climate crisis presents higher education worldwide with an enormous and complex challenge in addition to those other threats and problems. Higher education worldwide cannot avoid encountering the Anthropocene in a very material sense. This makes the matter urgent for academia. Colleges, universities, and other organizations working in the postsecondary education space will confront a mix of these challenges, although the climate may loom larger than all, depending on how global warming plays out and how civilization responds. In fact, the climate crisis might be the most dangerous threat academia faces in the twenty-first century.

Academia's exposure to climate harm in its primary form is the concern of this book's first chapter, which examines the many ways a college or university campus physically confronts the climate threat. From architecture to IT departments, we break down institutional exposures as well as ways academia can innovate.

That discussion is based on a certain assumption about likely ways climate change can play out, and this deserves some explanation. For most of this book, we will assume a scenario with a moderate amount of global warming and attending effects. It's based on the IPCC's 2021 SSP1-1.6 and SSP2-4.5 models (SSP stands for Shared Socioeconomic Pathway). That means, roughly, expecting a 1.5 degrees Celsius temperature rise over industrialism's baseline by the year 2030 and 2 degrees by 2050, heading over 2 by 2100, incidentally bypassing the 2015 Paris Agreement's 2 degree goal. In this scenario, the average amount of carbon dioxide in the world's atmosphere hits 450 parts per million (ppm) in 2040 and 500 by 2070. (For historical perspective, atmospheric CO_2 stands close to 410 ppm in 2022. It circled around 270 before the Industrial Revolution. Its first accurate measurement was 316 in 1958.) We are assuming the world hits either one major tipping point or none by 2100—

that is, either the West Antarctic Ice Sheet collapses into the sea, most of Greenland's ice sloughs off, the North Atlantic thermohaline circulation slows down majorly, the Amazon rainforest dries out or otherwise fails, or previously sequestered methane in northern Asia and North America escapes at scale . . . or none of those occur. Sea levels could rise by thirty to fifty centimeters by midcentury and reach as high as two meters higher by century's end.[8]

(It is possible that this model of global warming is either too optimistic or pessimistic. We will treat those two options in chapter 6.)

So far, we have spoken of academia as subjected to the climate crisis, as something to be acted on and actively endangered. What does reversing that construction yield? To return to our second question, what does academia have to offer the rest of the world as the entirety of civilization grapples with the deepening climate crisis? Why do colleges and universities matter to the global climate debate? This book's second chapter presents one classic answer in the form of academia's research mission. Historically, colleges and universities to varying degrees have produced and refined new knowledge. The mechanisms of inquiry, experiment, literature review, peer review, publication, and more have yielded a tremendous addition to our civilization's understanding of the world. A significant—and growing—swath of that knowledge concerns the climate crisis. We will see that there is a cluster of scientific disciplines already leading the investigative way, from meteorology and hydrology to computer science and, of course, climate science. We will also see other disciplines from not only the natural sciences but also the social sciences, humanities, and arts participating in the creation of knowledge concerning the changes wrought by the deepening Anthropocene. As the great topic mobilizes research within so many disciplines, it also crosses their frontiers in the form of growing interdisciplinary work. There are many ways for institutions to support this work. And the work will become more challenging as the climate crisis ratchets up in intensity.

Granted, universities and colleges do important research on climate change, and they also have a significant global footprint. We have another reason to talk about academia in global warming debates, and it is the biggest one so far for many institutions: the teaching role of higher education. Chapter 3 envisions how curricula—*what* we teach—may change as we advance further into the Anthropocene. To some extent, the expansion of climate research across the disciplines will drive a parallel growth across curricula. The chapter continues by examining how pedagogy—*how* we teach—might change as the climate crisis becomes increasingly important in classrooms and the world in general. It also addresses how *whom* colleges and universities teach may change, including institutions deciding if they want to teach and otherwise support climate refugees.

None of this teaching, research, or campus transformation occurs in a vacuum, and that leads us to our next two chapters, each of which connects academia to the world. Chapter 4 starts with local politics by taking on the old topic of how a campus interacts with its immediate community, which American culture memorably dubs "town-gown relations." The adjacent "town" (which can also be a rural district, a suburb, or part of a city) faces material threats similar to those of universities. That nonacademic community might be something for the academic community to research or to partner with. Changes to the town can ripple across the campus, including its businesses, nonprofits, primary and secondary schools, and government. Multiple tensions can crackle between the two, along with opportunities for productive collaboration. Their interdependence may deepen as the climate crisis develops.

Chapter 5 continues this theme of academic institutions and the outside world by widening its horizon to include how academia interacts

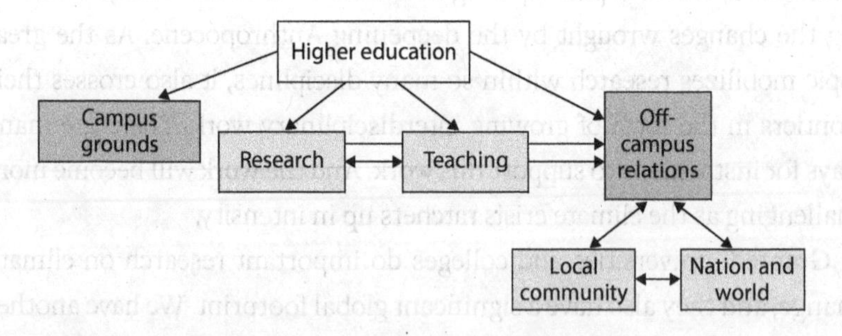

How higher education may interact with the climate crisis

with regions, nations, and the international community. This chapter begins by examining interactions with national governments and international structures, then addresses corporations and civil society before exploring potential sociocultural trends through which colleges and universities engage the Anthropocene.

With all of that ground established, the next two chapters extend our forecast for the next two generations by considering the future from different perspectives. Chapter 6 focuses on the problem of varying projections for the climate crisis and outlines two extreme possibilities. While the preceding text proceeded on the assumption that global temperature would rise by roughly 2 degrees Celsius by 2100, we now consider a more optimistic path, one where the world warms by only around 1 degree. We explore by what means this could occur and what it might mean for academia. Then we complement this by heading drastically in the opposite direction, envisioning a worst-case scenario. Once more, we outline how this could occur and what such a world could mean for colleges and universities.

Our final chapter is devoted to practical options for postsecondary institutions and individual academics as they consider the rest of the century under the aegis of the climate crisis. It starts with smaller options to better prepare academia for the Anthropocene's many dangers in the near term, then works up to the largest strategies for the rest of the century. Ultimately, we confront several possibilities. First, that academic institutions will transform themselves. Second, that outside forces will impose change on higher education. Third, that academia worldwide will intervene in the Anthropocene and seek to influence how humanity progresses through the next eight decades.

While each chapter approaches the intersection of academia and the climate crisis from a particular angle, there are many ways each angle overlaps with or connects with others. Some of the campus projects and designs described in chapter 1 return as inspirations in chapter 7. Climate threats described in this chapter recur in chapters 1 and 3. The worst-case scenario in chapter 7 may motivate decision-makers in ways we discuss elsewhere in the book.

Who Should Read This Book?

Universities on Fire is for anyone interested in higher education or climate change. The former audience can include college students, climate activists, parents of college-age children, university chancellors, government officials, businesspeople working in either academic or climate spaces, and the simply curious. Accordingly, I have tried to avoid jargon and have also assumed the reader is not intimately familiar with the Byzantine details of how the planetary higher education ecosystem works. I will, however, presume the reader is familiar with the basics of climate change. We will not have time to sketch out the basic science in the present volume. Helpfully, there are plenty of introductions and guides now available in multiple media.[9]

A Note on Method

As the preceding paragraphs have demonstrated, the purpose of this book is to outline and explore how the unfolding climate crisis can impact colleges and universities for the rest of this century. It is, by necessity, a futures work, drawing on decades of futures research. We will use several methods from that field, namely trends analysis (examining the present and recent history for powerful, shaping forces and extending them into the future) and scenarios (creating visions of the future based on the outcome of one or several trends). The climate crisis is, in a sense, the single gigantic trend shaping the world we focus on here. Underneath that header, we will identify trends resulting from that macro event. Climate change is so enormous a subject that simply grasping it requires new intellectual categories, like Timothy Morton's hyperobject theory.[10] In the present work, we will focus on the climate crisis as a driver of change. From that, will we derive scenarios of differing lengths.

Futures work occupies a curious position as of this writing. On the one hand, it is very much in demand, especially by corporations and governments, and especially as the rest of the twenty-first century appears more volatile, uncertain, chaotic, and ambiguous, or VUCA. (The term emerges from military discourse. I'm fond of Jamais Cascio's effort to create a successor acronym in case VUCA falls short.)[11] On the other hand, some hold futures analysis to be unserious or flat-out goofy. I suspect part of this stems from reactions to pop futurism as well as to stories of predictions that fail badly. Invoking the futures or forecasting field can run the risk of encountering suspicion or derision, based on a given reader's experience with poor practitioners, bad forecasts, or a sense that such work lacks rigor. In reality, professional futurists tend to emphasize not so much predictions but instead ways of thinking more strategically, thoughtfully, and creatively about multiple futures. We also tend to watch our forecasts carefully and to correct our present work against any misfires.

And misfires about climate change there will be. The world is getting stranger and more unsettled. Trends become unstable. The science evolves. As Stephen Nash put it, "We are moving out of the patterns to which we're accustomed and into variations that will be new, at times radical, and not as predictable. We don't know where they will ultimately lead until a new climate equilibrium arrives, perhaps centuries from now or longer."[12]

Yet we need to take futuring seriously, even essentially, in order to fully explore the possibilities of the climate change–higher education relationships. This is not only in the simple sense that we are talking literally about the future, and therefore futures thinking is apposite. Debates over the value of futures as a profession might mean less in the climate change field since the latter seems to be adopting the former to some serious degree. Futuring is the increasingly familiar general header under which much climate change research proceeds. The Anthropocene threatens to warp our inherited expectations of the future and our understanding of how the world works. In the words of one biologist with which we began this book, "A future is coming when nature is no longer fully natural." Writer and activist Bill McKibben frequently argues for the sheer difference of the decades to come, even titling one book with a new name for Earth to make that point. Thinking and planning about the Anthropocene as if the future will simply continue the past is a recipe for disappointment and disaster.[13]

Trends analysis has to take care to avoid that trap. In this book, we will draw on the most recent research and modeling to obtain the best available insights into the unfolding, strange future. We will also cross the disciplinary streams by working across intellectual domains, from the natural sciences to the social sciences, humanities, and creative arts, trying to keep the windows of possibility open for readers. All of our discussions are grounded in basic data, especially when it comes to higher education—yet while keeping an eye on how sudden developments and changes can skew trends in weird directions. We necessarily combine evidence with speculation.

Scenarios play a somewhat different role. They are audience participation exercises of a sort, short narratives aimed at inviting readers to imagine themselves in different years to come. Accordingly, their details are more sensory and less data based, although each scenario stems from trends backed up by evidence. Consider them strategic windows through which you can glimpse a possible future, and then start thinking about how to traverse the distance between here and there—either how best to reach a desired future, or how to avoid a negative one.

The reader may note that these trends and scenarios are plural, not singular forms. This text, like most futures work, addresses the future as a set of possibilities rather than a single, fixed, and unitary production. Similarly, this book does not restrict itself to one model for understanding how human society as a whole may respond to the deepening climate crisis. It may be that our species comes to a unifying consensus about moving beyond the industrial era of world-altering greenhouse gas emissions and decides to follow an orderly and humane transition into a new epoch. Recent history—well, all of recorded history—suggests this will not be the case. Instead, we are more likely to realize multiple scenarios at the same time around the world. Some higher education systems will follow one source while others will seek another. The societies within which academic institutions are embedded will also most likely pursue divergent courses.

Supporting these scenarios and our exploration of their implications is a body of research. Readers will note that this book's sources are heterogeneous in many ways. In publishing terms, they include peer-reviewed scholarly materials, preprint articles, journalism, professional organizations' websites, video clips, podcasts, and books. This reflects the location of the best information about the many topics under discussion here. Some events and arguments, for example, are moving too quickly to be captured in print and hence can best be captured in social media and preprints. Traditional scholarly articles represent more established research directions. In contrast, some models and schools of thought appear in the greater length allowed by monographs and other books. At times, I have cited secondary sources in order to make them more accessible to nonspecialist readers; they point to primary materials for those who wish to probe more deeply. In terms of content and authorship, most of the evidence cited in this book is from scholars, followed by commentators, including journalists, along with a touch of literary fiction. This reflects the range of people thinking about academia in this age of transformation, biased toward professional academics. I have tried to draw on scholarship across the disciplines in a balanced way, seeking to avoid the bias of my own education and some of my previous research, which was centered firmly in the humanities. I have also interviewed some scholars, administrators, and commentators, and those conversations are indicated in the notes at the end of the book.

At the same time, futures work eventually runs out of recent history and present evidence, heading into the realm of informed forecasting. In those instances, the text builds on what has been established so far, then extended with the tools of futuring.

What This Book Covers and What It Does Not

In terms of time, the present book seeks to explore roughly the next eight decades of higher education, with a timeline running to the year 2100. Put another way, that stretches over the next two to three generations of academia. It is also based on the present day and some recent history.

For some historical context, that futures ambit will cover roughly the 47th and 48th generations in the lifespan of the oldest university, the University of al-Qarawiyyin (جامعة القرويين), which began in the ninth century. It would be the 37th and 38th generations who experience the University of Bologna, another popular candidate for the world's oldest university (founded 1088), and approximately the same for Oxford University, assuming all three survive this century. On a smaller historical footing, getting to 2100 takes us through the 13th and 14th generations of American higher education.[14]

Our social focus is on the higher education sector as a whole. We will necessarily address academic institutions, their features and dynamics, and their strategic options and operational issues. This includes all types of colleges, universities, and similar entities. Public and private, online and off-, research intensive and teaching focused are within our scope.

Geographically, this book touches on postsecondary education worldwide, as the climate crisis is one that refuses to be contained by national boundaries. Examples appear on every continent and stretch across the nations, from Iran and New Zealand to Uganda, Britain, the United States, and China. I must confess that that ambit is enormous and will run into limitations of space. It will necessarily leave out many institutions and examples not from ignorance or criticism but in order to keep the text within the covers of a single book. It is my hope that this volume accelerates research into academia's Anthropocene and that other, more gifted scholars can add to this limited sally.

So much for what the present volume seeks to cover. We should also be clear about what this book does *not* address at length due to limitations of space and time. To begin with, it does not offer a Climate Change 101 class. We are assuming the reader is familiar with the basic science describing climate change or is willing to partake of the many resources available to serve this need. We will not sketch out the history of climate change or its scientific apprehension. For example, readers should expect references throughout to carbon dioxide and methane as greenhouse gases without being informed how greenhouse gases retain solar heat.

Our focus on higher education means paying less attention to other institutions. We will touch on other educational and education-related sectors to an extent (primary and secondary schools, museums, libraries, archives, etc.), but postsecondary education remains our primary object. We will bring in governments, businesses, and nonprofits as they impact colleges and universities, but will not investigate them deeply on their own.

This book also does not offer a comprehensive survey of other challenges higher education confronts. That was the function of its predecessor, *Academia Next* (2020). That volume examined a broad range of non-Anthropocenic trends reshaping higher education, including developments outside academia (demographics, macroeconomics, public attitudes toward higher education, and quite a bit on digital technology) as well as those within (enrollments, finance, internationalization, adjunctification of the professoriate, and so on), while in this text we remain strictly within the bounds of a single, if immense, driving force. Another difference is that *Academia Next* focused on American colleges and universities, while the present book addresses higher education worldwide. Yet both offer perspectives on the future of academia. The climate crisis intersects with the trends *Academia Next* outlined in many, many ways. Please assume that those trends remain in play as you follow these pages. For example, the previous book identified the demographic trend of decreasing childbirth as an issue for higher education enrollment, especially for institutions that focus on teaching traditional-age undergraduates. Climate change may deepen that trend in one way, as rising numbers of people consider not having children due to fears of the worsening climate crisis ahead. In this sense, the reader can understand that this volume can be understood as a sequel to *Academia Next.* Reading the first book will add to the reader's takeaways from the second.[15]

The Role of COVID-19

In *Academia Next,* a wide range of threats appeared from a range of trends. Among others, the book raised the specter of a global pandemic and how such a disaster might impact the academy.[16] In my experience, no reader found this assertion remarkable until the COVID-19 virus swept out of Hubei province and attained pandemic status worldwide. Given the depth and breadth of the resulting public health crisis, we must address that grim topic here. COVID will also appear throughout this book.

Considered in relation to climate change, we can derive lessons from the human and the academic experience of the pandemic. To begin with, the deep and costly shock many nations experienced may make more communities take disaster planning more seriously. Tōhoku University professor Takako Izumi found that many universities underprepared for COVID, although those with hard experience of SARS may have been better positioned for resilience.[17]

Viewed in another light, some may view COVID as a kind of dry run for the human response to climate change. Bruno Latour floats this idea, which seems plausible, as both events involve civilization responding to a global crisis involving deep engagement with bleeding-edge science. The pandemic did demonstrate humanity deliberately reducing carbon emissions, including from college campuses, during lockdowns and quarantines. If we make this comparison, we may deduce that humanity is likely to politicize climate science even more than it does now, based on how thoroughly many nations did so with various aspects of COVID: its origin, identity and deployment of vaccines, contradictory advice from experts, and so on. Nation-states and localities often took the lead in mounting responses, while international cooperation seems relatively slight in comparison. We may be in store for what Geoff Mann and Joel Wainwright have dubbed "Climate Behemoth," when many nation states individually take the lead in climate policy, lacking coordination and often acting in some form of denial.[18]

As COVID casualties mount, the comparison between virus and climate crisis may elicit pessimistic conclusions. Some, such as Andreas Malm, see the pandemic setting climate activism back:

> We have to be honest about the situation we find ourselves in. COVID-19 has brought about the sudden obliteration of the climate justice movement in terms of everything that had been built up by the end of 2019. Since early 2020,

> COVID-19 has completely paralyzed all the most promising developments in the environmental movement—Fridays for Future, Extinction Rebellion, Ende Gelände, and so on—this is a situation of grave disaster. Prior to this, there had been a growing momentum toward aggressively disrupting business as usual.[19]

In 2021, the Bulletin of the Atomic Scientists surveyed the landscape and determined that "the pandemic serves as a historic wake-up call, a vivid illustration that national governments and international organizations are unprepared to manage nuclear weapons and climate change, which currently pose existential threats to humanity." It is an understandable finding, given the pandemic's death toll—a mind-bending 6,350,899 as of July 10, 2022, by one leading count, which is likely an undercount—and the still larger number of people suffering chronically from long COVID. If the pandemic was a dry run, humanity showed many, many weaknesses.[20]

Academia also revealed its own weaknesses, depending on one's perspective. It is difficult to determine how many people within the academic world were infected, injured, killed, or afflicted with long COVID, as institutional data is not always consistent and comparable. The *New York Times* devoted dozens of reporters to the problem and barely managed to build a tracker for half of American colleges and universities.[21] We know that some of our community did get infected. Some were injured. Others suffered long COVID, and others still died. It is plausible that those numbers would have been lower if more institutions chose wholly online learning rather than some form of in-person education. Interactions between campuses and communities were uneven, including universities being ordered to change course by local authorities, charging them with spreading the virus.

There is no sign that the climate crisis will be a more lenient challenge. Instead, there is now a crying need for academia to help civilization through our research, our teaching, our community involvement, and our public roles.

Yet the dress rehearsal might be a flawed model for comparing virus and hyperobject. After summoning it, Latour argues against the idea, finding that the essential nature of each crisis differs by the respective roles of humanity:

> In the health crisis, it may be true that humans as a whole are "fighting" against viruses—even if they [viruses] have no interest in us and go their way from throat to throat killing us without meaning to. The situation is tragically reversed in ecological change: this time, the pathogen whose terrible virulence has changed the living conditions of all the inhabitants of the planet is not the virus at all, it is humanity!

Perhaps one common thread between COVID and the climate crisis is an ecological one. The succeeding threats could cause civilization and academia to rethink humanity's relationship to the nonhuman world. Latour offers this conclusion in a more optimistic vein than the Atomic Scientists':

> What allows the two crises to occur in succession is the sudden and painful realization that the classical definition of society—humans among themselves—makes no sense. The state of society depends at every moment on the associations between many actors, most of whom do not have human forms. This is true of microbes—as we have known since Pasteur—but also of the internet, the law, the organization of hospitals, the logistics of the state, as well as the climate.

Academia is well positioned to make that connection through research, teaching, and relationships with the broader world.

Notes on Language Used in This Book

I'd like to conclude this introduction with some words on language. First, this book will use three terms more or less interchangeably and with some unfairness: climate crisis, climate change, and global warming. "Climate crisis" is currently the term most scientists and advocates favor, given its implications of warning and action. I will also use "climate emergency," which has a similar meaning and resonance, although less often, as it is not so widely recognized. "Climate change" is a widely recognized expression, easily understood, and so it will appear throughout this volume. It lacks the pungency of "climate crisis," since "change" is a neutral, rather than frightening, term, but its currency remains broad. "Global warming" is an older phrase, dating back to the 1980s, and on the face of it serves us well, describing the planet's heating. It is also popular, although its star may be descending, at least in the world of books, as Google's Ngram, a massive digital database of scanned books, shows.

Some scientists, activists, commentators, and others refer to our time as the Anthropocene, identifying the modern era by arguing human-caused climate change amounts to a new period in geological history. There are different ways to date the Anthropocene's beginning, but for our purposes, it begins with the first Industrial Revolution and humanity commencing to emit greenhouse gases at scale. The term "Anthropocene" has many advantages, such as making us realize the scope of anthropogenic warming and also the extent to which its impacts are of a long duration. For those reasons, we will rely on it in this text.

I am personally fond of the expression "global weirding," coined by the Rocky Mountain Institute's Hunter Lovins, as it amply describes the chaotic nature of the crisis, as well as how humanity finds itself estranged from the earth. It makes room for the wild range of climate variation the Anthropocene entails, from superheated northern Siberia to frozen Texas, both droughts and floods. I like the way it opens up possibilities

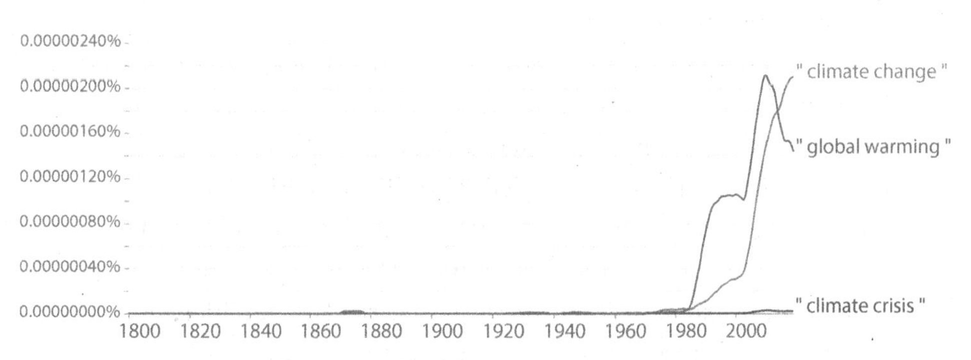

Google Ngram Viewer, comparing English-language book mentions of "climate change," "climate crisis," and "global warming" over time. Google Books Ngram Viewer, http://books.google.com/ngrams

while making us more cautious. Unfortunately, it remains a rare term, so I will restrain its use in these pages to some degree, while hoping its occasional appearance intrigues readers enough for a reconsideration.[22]

All of that said, the language around climate change is likely to mutate over the rest of this century, as language often does. By 2030, we may consider "climate crisis" to be cringeworthy and have been heartily succeeded by a new term. By 2075, "global warming" may come back as a deliciously retro term, while 2030's phrase, whatever it turns out to be, will be a historical curiosity.

In the chapters that follow, we will frequently pair two phrases that some oppose today, which deserves some explanation. "Climate mitigation" refers to efforts to slow global warming, such as reducing greenhouse gas emissions or planting massive numbers of trees to soak up carbon dioxide. It can include installing direct air capture (DAC) devices at scale. This technology draws carbon dioxide directly out of the atmosphere and sequesters it elsewhere. One of the more controversial mitigation ideas is geoengineering, large-scale physical efforts to alter the Earth's climate, ranging from planting millions of trees to slow a desert's spread to orbiting mirrors in space to reflect some of the sun's incoming light. In contrast, "climate adaptation" denotes ways by which we change behaviors to adjust to a new environment. These can include migration, acclimating to higher temperatures, building seawalls, genetically engineering species for transformed habitats, or altering diets to account for altered food systems.

The differences between climate mitigation and adaptation are clear, but sometimes we will use them interchangeably when discussing institutional strategies. That is, when a university faces investment choices over decades, its options may include both climate mitigation and adaptation. Its researchers investigate both responses to the crisis, and both movements require institutional support. The same goes for teaching, as adaptation and mitigation appear in a curriculum. Additionally, colleges and universities may pursue changes to their physical environments that meet both mitigation and adaptation needs. For example, buildings providing their own water (from rain, transpiration, etc.) reduces strains on local aquifers, which helps reduce water problems while also preparing for a world where some of those hydrological issues persist.

About the present volume's title: the word "fire" in *Universities on Fire* has four different meanings. First, it describes the literal burning of academic property caused by climate change, especially on campuses exposed to combustive threats: those in drylands, with poorly defended electrical infrastructure, or containing or adjacent to flammable materials. Obviously, fire is not the only way our institutions may be harmed, as this book will show. In the title, fire stands metonymically for the gamut of possible damages and injuries to property and people.

Fire also refers, by analogy, to activism and excitement. A university may be on fire with the organizing energy of its students, not to mention its faculty or staff. New programs may elicit enthusiasm from students. Climate research can light up the crisis and energize the world. The Sunrise Movement, itself named after our fiery star, is conducting a Generation on Fire mobilization and march across the United States as I write this.

Fire can be a destructive force, of course, and that implies a third sense of the word here. Worldwide, some colleges and universities are under threat for reasons other than climate change. Multiple national and local political forces impinge on academic freedom, student learning, and institutional sustainability. War, disease, and economics also hit postsecondary education. We should assume those threats will shift, mutate, and persist for the rest of the century. New threats may appear as well, such as negative side effects of new religious movements or some technological fallout. Colleges and universities planning for or reacting to the climate crisis may scramble to do so when these other threats attack them. Readers should bear this context in mind during the following chapters.

In the United States, non-Anthropocenic threats take a particular cast, which was the subject of my previous book. Higher education capacity seems to be overbuilt for purpose, as enrollments have declined for a decade. Institutions have massively relied on temporary, or "adjunct," faculty, in part to cut costs. Some have also cut tenure-track faculty, especially in the humanities, a process I have dubbed with a chess term, "the queen sacrifice." State governments have reduced their financial support to the majority of American universities, driving campuses to raise prices and students to take out ever-increasing loans. The financialization of funding has led to an extraordinary bolus of student debt, a crisis that has already negatively impacted some students' finances and mental health. Elite institutions are secure from these threats, protected by the combination of wealth and reputation, but among the majority of American institutions are many that can hear the crackling of flames.

Finally, the present volume's title refers to fire in a literary or mythic sense. In folklore, literature, religion, and psychology, fire sometimes means illumination, both literally and symbolically. Fire destroys, but it also shines a light. We often describe insights or realizations in terms of seeing the light or being enlightened. The root of illumination is ultimately fire. Colleges and universities are based on enlightening students through teaching, as well as expanding humanity's illumination of the universe through research. One of our essential roles in a world on fire is to help us see the crisis more clearly. The world deeply needs higher education to help it through the Anthropocene, and academia needs to heed climate change for the rest of this century.[23]

I hope this book can provide some illumination as we look ahead, as well as a framework for readers to create their own knowledge.

On a personal note, I started working on this book during a time when evidence for climate change was rising and so was the political controversy about how to respond. I live in the United States. Donald J. Trump was president when I began writing this book, and Joe Biden succeeded him as I completed the manuscript, a pairing that neatly illustrates the broad range of political views of the subject.

As I wrote, world temperatures reached some of the hottest on record. "July is typically the world's warmest month of the year, but July 2021 outdid itself as the hottest July and month ever recorded," according to the United States National Oceanic and Atmospheric Administration's administrator. Brutal heat waves and fires roared across the western United States, so intense that *Scientific American* described them as entering "uncharted territory." Lake Mead, a crucial water source for the American West, dried out to less than one-quarter its normal size, leading to its first federal shortage declaration. An Alaskan town has started evacuating because of icequakes, caused by ice melting at an unusually rapid pace, that are rocking the state. Animals are now *observably changing size* in response to changing environments.[24]

The climate crisis didn't only manifest so starkly in America, of course. Northern Siberian temperatures reached nearly 120 degrees Fahrenheit, and vast wildfires hit that normally very cold region. Scandinavia endured unusual heat waves. Unusually large fires broke out in Italy and southern Turkey. South Africa experienced drought-driven water shortages. Floods caused deaths and property damage in western Germany and in central China. Greenland's ice sheet melted at a faster pace than expected, due to a heat wave. Rain—rain!—fell on that ice-covered land. Meanwhile, the vast Atlantic Meridional Overturning Circulation (AMOC) system, which carries warm waters and air from the Caribbean up North America's East Coast and then into northwest Europe, showed disturbing signs of a potential slowdown. Globally, the atmospheric CO_2 level is now around 419 ppm, a record for human history.[25]

Humanity wasn't idle in the midst of these events. The responses I saw rising about me as I wrote were numerous, energetic, and suggestive of the first stage of this new era. European activists took Norway's largest oil company to court to reduce its drilling. The United States National Weather Service abandoned a weather station on the Atlantic coast, determining that sea level rise and growing storms made it untenable. The first direct air capture facility started capturing carbon in Iceland. An Indonesian coal mining company announced it would build solar plants literally on the sites of former coal mines. At the same time, I tracked human developments of a very different slant. Governments in Australia, Norway, and Britain decided to support new carbon-burning projects, including setting up new coal plants and offshore oil drilling platforms. Britain and China continued burning large amounts of coal. India used so much coal in 2021 that the country began importing more of the fuel. An American court sentenced a climate activist to eight years in prison for damaging oil company property. To paraphrase Christopher Marlowe, this is the Anthropocene, nor are we out of it.[26]

Writing this book during this time meant risking extreme mood swings. At times, the sheer amount of creativity, innovation, and interest from academics gladdened me. The capacity of academia and the world for change is breathtaking. At the same time, the potential for disaster—and its first realization—is a terrible thing to live with. Chapter 6 took me to extremes of optimism and horror. The latter experience reminds me of what Warren Ellis once dubbed "abyss gaze," a mental condition futurists suffer from when they stare too long into the worst futures.[27] I hope I didn't inflict too much of that on my family and friends. The range of negative reactions often entailed in contemplating the worst options—despair, gloom, bitterness—may well attend your own reactions, reader. I cannot apologize for this.

Yet it is not by myself, but only through networks and relationships that a work like this can be accomplished. I tried to thank all in the acknowledgments, but here I wanted to emphasize that theme. One person can try to grapple with a hyperobject like climate change. It is much easier to do so with kindred spirits. The results are better as well. During the past few years, public discourse has in some registers become more bitter and difficult, if not shut down in favor of shouting at the enemy and rallying one's perceived troops. Discussions are badly needed, sometimes hard to host, and remain so vital for our collective life.

To all of those people who waded into those discussions and participated in our networks together, who contributed to the thinking and research of this book—you have my deepest appreciation. The book's strengths are yours; any flaws are solely my own.

I UNIVERSITIES ON FIRE, UNDER WATER

CHAPTER 1

Uprooting the Campus

It was mostly about how we can keep Eckerd [College], Eckerd while still considering what climate change is . . . It's apparent, it's coming and how we're going to conquer that battle while still keeping Eckerd the same Eckerd that we all know and love.

—UNDERGRADUATE STUDENT ELLI ROGERS

From the beginning of American history to 1889, the main building—often the only building—of no fewer than 62 institutions of higher learning burned to the ground.

—VIRGINIA SAPIRO

We can start with a challenge. Climate change will literally reshape many institutions for the rest of the twenty-first century. The physical campus confronts the climate crisis unavoidably and obviously by the sheer fact of being embedded in the physical, ecological world. Academic buildings and libraries, lawns and streams, meals served and plants raised, cars and shuttles, researchers and students are subjected to storms, heat, water, sand, and the manifold other components of climate damage. Looking ahead, we may have to rethink nearly every aspect of each brick-and-mortar academic institution in terms of vulnerability to environmental stresses during the worsening climate crisis. At the same time, we should rethink the opposite: how campuses as they presently exist might contribute in small or symbolic ways to either worsening or addressing climate change.

Facing this dual challenge of adaptation and mitigation means bearing in mind the many distinct yet interconnected elements of a physical campus facing the Anthropocene. These elements will vary from campus to campus, of course, based on individual circumstances—that is, not every college has residence halls or a museum, lazy river, dock, prison, or on-site power plant. A campus may sprawl across a county or confine itself to a single building. It may present itself as the acme of technological progress or a richly conservative temple of tradition. The sheer variety of academic entities around the world is testimony to our ability to be innovative and creative—and to make those ideas work. At the same time, there may be more to the physical campus than we normally think, depending on our perspective.

We can start with the physical plant, the instances of the built environment that occur on academic grounds. Academic buildings loom largest, containing classrooms small (seminar) and large (lecture hall), laboratories, faculty and staff offices, their administrative units, and so on. Other buildings nominally focused on different aspects of the academic mission may also be present: libraries, clinics and hospitals, performing arts centers, archives, museums, dining halls, observatories. For students living on campus, there are dormitories or residence halls. Associated with those, there may be many other structures, depending on an institution's commitment to student life and care: wellness centers, stadiums, student clubs, student unions, climbing walls, gyms, barracks, cemeteries, and more. Some campuses host third-party businesses that supply food, office supplies, books, or clothing. Supporting these buildings are other, not so well-known structures, such as those housing campus vehicles, electrical power storage and transmission, steam heat, security forces, and maintenance tools. An infrastructure connects all of these to one another, often underground or otherwise concealed: power lines, ethernet cables, water pipes, sewage pipes, steam tunnels. These buildings may not be fully occupied by the academic community; indeed, some might not be owned by a college or university but instead leased from a landlord. And those structures may occupy only a portion of campus grounds (some buildings and spaces may stand on other land). The rest are devoted to lawns, parking lots, sidewalks, stands of trees, public art, pedestrian bridges, arboretums, creeks, ponds, beaches, trails, and malls. With just this brief sketch, the picture emerges of even the smallest campus as akin in size and structural complexity to a town, and the largest to a small city.

Universities' and colleges' holdings are not necessarily limited by their campus boundaries. Some own other physical properties beyond their primary site. Adjacent to that site, they may possess a range of buildings in the immediate and surrounding community, from offices to student housing, farms, and industrial sites. They can also include properties to be developed in the future. Beyond the local community, institutional holdings can dot the world. Some have branch campuses in the same province or in other countries. They can support research sites at remote locations. Others have land assets that act as investment properties. Campuses have been known to own golf courses, ski resorts, trails, and other recreational holdings. Those universities offering study abroad programs may lease or own properties in other nations and on other continents.

Each physical piece of these academic enterprises, from a power line connecting a science building to off-campus electricity to a lab perched on the edge of a rising sea, is potentially vulnerable to climate change. Taken together, they constitute the elements of comprehensive, institutional exposure.

How and where might they be struck?

Campuses located on seacoasts, floodplains, or along rivers can face a variety of hydrological threats. As Lee Gardner reminds us, "Some institutions find themselves facing '100 year' floods every few years." Storm surges can send water across campus grounds and into the lowest levels of buildings and infrastructure alike. More attenuated flooding can not only damage property but spread rot and the basis of diseases. More drastically still, sea level rise can gnaw away campus grounds and buildings, ultimately overrunning an entire college or university. Think, for example, of the roughly forty academic institutions in West Bengal, facing the Indian Ocean. Or consider Maldives National University and nearby colleges, as *all* of that island nation's lands confront the likelihood of being completely submerged. The Mekong delta—low lying, rich with rivers, and exposed to the Pacific Ocean—contains several academic institutions, including Tra Vinh and Can Tho Universities. Other examples include campuses along the Egyptian Nile, on the coast of Niger, along China's Pacific coast, and the many American colleges and universities on the Gulf of Mexico or southern Atlantic coasts.[1]

While these damages can be severe, we may anticipate additional damages through the secondary impacts of water. Indeed, some are already apparent in some locations. Salt water can not only damage buildings and lives but also infiltrate freshwater by entering aquifers, which can kill, and also replace, plant life. Lakes on or adjacent to campus

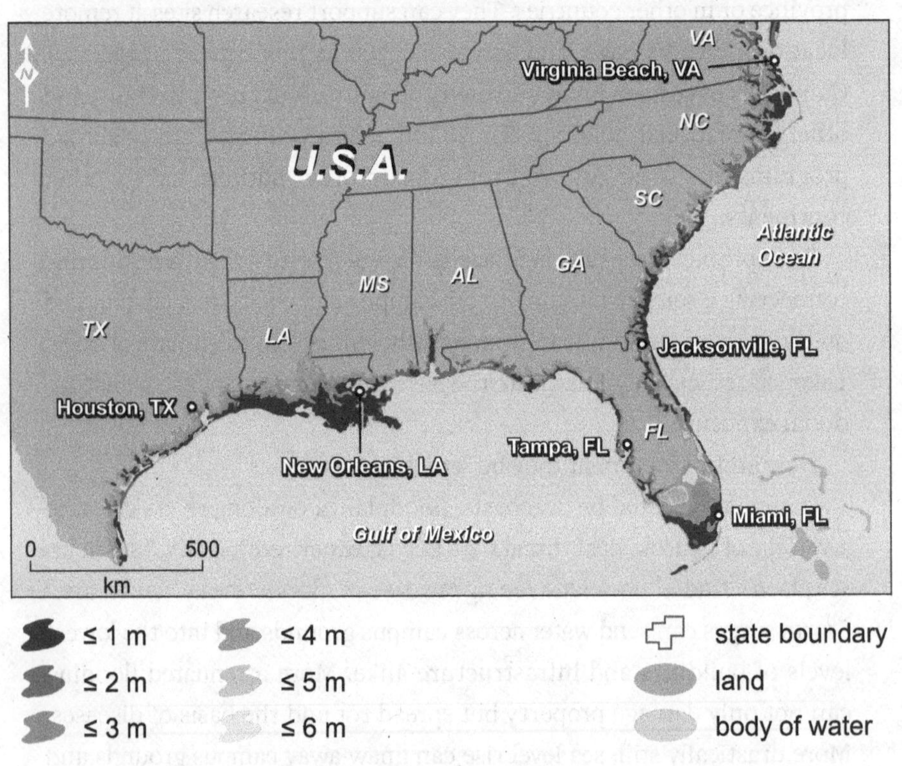

Potential sea level rise around the American Southeast. Reprinted by permission from Springer Nature: Jeremy L. Weiss, Jonathan T. Overpeck, and Ben Strauss, "Implications of Recent Sea Level Rise Science for Low-Elevation Areas in Coastal Cities of the Conterminous U.S.A.," *Climatic Change* 105, no. 3 (2011): figure 1. Copyright © 2011

grounds are likely to suffer decreasing oxygen levels, due in part to increasing temperature. This can kill off fish, turn clear waters cloudy, and also release buried methane. Early evidence of this kind of process appeared in a recent Norwegian report, which found lakes in that country already starting to change color. Tree lines have moved, snow seasons have shrunk, precipitation has increased, glaciers have shrunk, and animals have relocated. Landslides, dam failures, building collapses, and surprise flooding have occurred. All were either caused by climate change or through processes that climate change exacerbates. All of these alterations can impact campuses. Some will have to install or expand water purification systems, expand sewage systems, erect barriers against potential floods, or re-regulate swimming areas. Others may rethink their identity as their physical profile changes, especially institutions where natural beauty and the natural setting play a key role. For a vivid example, consider University of Cumbria in Ambleside, located in England's greatly storied Lake District. That situation looms large in the university's outreach and image. What happens to those representations when the reality changes into a new form, far from what it was in the nineteenth century?[2]

Water changes do not act alone, of course. They are often accompanied and driven by rising temperatures worldwide, hence the still popular term "global warming." The hottest regions on Earth, such as southern central Asia, the North American Southwest, the coastal Middle East, and the Persian Gulf, may experience temperatures increasingly likely to debilitate, injure, and kill human beings. We need to distinguish here between two types of heat, very humid and very dry. High "wet-bulb" temperatures, a measurement that combines heat and humidity, can reach high enough levels to debilitate and injure humans in certain hot and damp areas, such as much of South Asia or along the American Mississippi River. At the most extreme, some regions will see wet-bulb temperatures exceed 37.5 degrees Celsius (99.5 degrees Fahrenheit), beyond which human bodies cannot regulate their heat and therefore start to become sick toward exhaustion and death. Additionally, the combination of increasing heat and humidity can facilitate the growth of mold within buildings, which can in turn worsen human health. Are campuses in and near such areas prepared to rely on heavy-duty air-conditioning, vapor barriers, dehumidifiers, and other mechanisms to keep their populations safe from this danger? How will academic communities respond to their members being sickened, injured, or killed by the transformed climate? How many colleges and universities, facing these threats, will decide to close or relocate their physical campus instead?[3]

The combination of water with heat leads to another problem. We are also likely to experience increasing rainfall in different parts of the world, adding downpours to floods far too often. That results from the way warming temperatures boost the amount of water the atmosphere can hold. Certain European areas or even the entire continent will likely receive more rainfall, especially as storms both intensify and increasingly stall in place, leading to more "quasi-stationary" storms, which spend more time pouring water down on the same spot. Increased flooding is a logical result.[4]

Alongside greater rain- and snowfall looms another danger: more intense storms. The Pacific and northern areas of Asia, northern Europe, and the Atlantic regions of North America are likely to receive increasing amounts of water in the form of more frequent and/or wetter storms. This can lead to floods, of course, which become dangerous when they exceed established safety measures designed for the pre-Anthropocene era. Monsoon areas around the world—southern Asia, western Africa, Malaysia—may experience downpours beyond their ability to cope. An individual campus can experience this threat simply in the form of greater amounts of storm damage to buildings, grounds, and population. Sudden floods can cause further injuries.

The climate crisis promises extremes of precipitation, and not just in terms of excessive downpours and humidity. The opposite of increasing water is intense dryness, which also harms humans. It can reduce the capacity of natural and artificial water supplies, causing further distress, disease, and death. Looking to 2050, one model projected decreasing water resources, rising costs, and growing water insecurity: "Results show that, next to the existing regions experiencing groundwater depletion (like India, Pakistan, Central Valley) new regions will develop, e.g. Southern Europe, the Middle East, and Africa . . . [W]e estimate that in 2050 groundwater becomes unattainable for ~20% of the global population, mainly in the developing countries and pumping cost will increase significantly."[5]

Campuses located in drylands worldwide or near deserts may face withering dryness, from the African Sahel to the American Southwest and southern Asia to much of the Mediterranean basin. Such aridity may increase in areas where global warming dries out winter snowpack, giving successive years even less water.[6]

Dry heat can strike populations unevenly based on social, cultural, and political factors. For example, higher temperatures are more likely to impact Hispanic and especially Black people in the United States due to economic, housing, and geographic differences:

> Between 1971 and 2000, US counties with more than 25% black residents endured an average of 18 days with temperatures above 100F (38C) compared to seven days per year for counties with fewer than 25% African Americans. By mid-century if Paris climate accord targets are not met, US counties with larger black populations will face a staggering 72 very hot days a year on average—compared with 36 days in counties with smaller African American populations, according to the [Union of Concerned Scientists].[7]

Rising temperatures impact not only humans but also plants and animals. Increased heat drives species to relocate to more congenial areas, which means changes in habitat and migration. This can reduce those populations, depending on their ability to adapt. It can also create ripple effects across local interspecies relationships as predators, prey, symbiotes, and various dependencies are rearranged. This causes two effects on the human world, starting with changing the biomes we inhabit. Changing temperatures impact plants, which can drive agricultural declines in areas that warm up and dry out, such as southern Europe and Taiwan. Increasing temperature causes prairies to advance, as in central Asia, South America's pampas, and America's upper Midwest.

Further, local and economic output can drop as soil quality deteriorates, water becomes scarcer, crops become more difficult to grow, and machines (including energy sources, such as solar panels and wind turbines) become more expensive to maintain. This in turn makes nonimpacted land more valuable, which can further marginalize the poor drylands—where roughly two billion people live, mostly in the developing world—that are especially vulnerable to desertification. Drying out vegetation makes fires more frequent and dangerous. Larger fires loft more ash and other particles into the atmosphere, which can degrade air quality in large areas, leading to breathing problems and even death. Further, as soils dry out and plant life declines, both land and vegetation lose the ability to sequester carbon—that is, desertification accelerates climate change, becoming a dangerous, vicious circle. We can already see signs of these changes in contemporary Brazil, which is suffering a terrible, supposedly once-in-a-century drought, or in Guatemala, similarly hit.[8]

These changes to the nonhuman biological world may have clear results on the human academy. To begin with, campus grounds change with rising heat as plants become unable to thrive in a new microclimate, which literally alters the face of an institution, either reducing the amount of flora that can flourish or replacing them with other species. Agricultural programs have to pivot to a new reality for their operations. Off campus, a dislocated agricultural sector can cause an economic recession or worse, complete with depressed businesses and an unhappy population.

More dangerously still, desertification is likely to expand in drylands around the world. Drylands already constitute roughly one-third of the earth's land surface and are quite vulnerable to becoming more arid still. Encroaching deserts can drive increasing amounts of sand and dust in surrounding areas, which cause breathing problems for humans and other animals along with damage to machines and buildings. Other forms of bad weather driven by changing climate can hit campuses. Rising average heat is troubling enough, but the extreme temperatures of heat waves can damage campus infrastructure (think of melting tarmac or electrical infrastructure) and populations. Increasing aridity makes fires much more likely as various flora and natural remains dry out and the volume of combustible material soars. Additionally, a campus doesn't need to have a fire occur on-site or directly nearby for air quality to suffer, as particulate matter can drift great distances. Academic populations may endure breathing problems dozens or even hundreds of miles from a conflagration.[9]

Pacific Union College offers a case study in how an institution can adapt to rising threats from fires. PUC has already experienced enough fires that the danger is close to routine. Its process for dealing with approaching flames starts with preparing its physical property. PUC now has a conservation easement and firebreaks between the campus building perimeter and the most likely fire path, a setup closely studied by its faculty. The college has established fire crisis relationships with the local fire department, the local forest manager, and the California Conservation Corps. Members of the university community are also prepared to play crisis roles, as some staff serve in the fire department and some students pursue undergraduate degrees in first response or in conservation technology. Hiring committees keep an eye out for job candidates who show resilience and an aptitude for handling crises. Internally, PUC has an incident command team structure in place, chaired by its chief financial officer, and including representation from departments of public safety, student affairs, information technology, and communications. The campus boasts a very small airport, which can be used to stage firefighting aircraft. When fires approach, the college activates its command team, evacuates most of its population, and helps keep the flames away from campus.[10]

Taken together, we see colleges and universities exposed to increasing damage as the climate crisis continues. Most harmful are the threats to human lives. The threats to a campus's built environment follow, including harm to beloved or historical structures as well as increasing maintenance costs. Beyond the financial level, how does a campus change when its lawns, forests, buildings, and waters become something new? When does its identity become something new, and perhaps not as attractive? At a different level, there is also the threat of reduced campus property values, which can increase costs of borrowing money for capital projects.[11]

Given such threats, how might colleges and universities respond? In anticipation, how will we plan? To answer these questions, we should add another motivation to our thinking in addition to self-preservation and protecting our communities. Academic institutions are not only exposed to potential damage but also contribute to some small degree to the process of global warming. To repeat an earlier theme, campuses are akin to towns and small cities. Like those settlements, campuses produce greenhouse gas emissions. Also, like towns and cities, campuses emit GHG in different ways depending on their composition. A college may purchase electricity to warm or cool its buildings from sources that burn coal or oil. Its faculty and staff travel for work via cars, trains, or aircraft, lofting more hydrocarbons into the warming atmosphere. Food service provides meats and other animal products, the production of which generates both methane and carbon dioxide. Carbon dioxide is heavily involved in the production of concrete, which forms some degree of the academic built environment.

When I have raised this point with academic audiences, pushback often occurs swiftly. The counterargument usually starts on a quantitative note, showing that a given college or university's GHG emissions are trivial on the world stage. Qualitative responses are also fairly popular, with academics arguing that these operations are necessary for human development and that they justify the comparatively meager amount of atmosphere loading. In particular, the contributions universities make to human progress during the Anthropocene are more important than their very small role in furthering the crisis.

My response to these good ripostes is to return to the analogy of towns and small cities. Would we grant a conurbation a carbon pass because its leading industry is solar panel manufacture? Should we not apply decarbonization policies to towns because, individually, they each represent such a small fraction of the Anthropocenic whole? Further, the counterarguments deprive two populations of agency on this score. They forget that some campus populations—faculty, staff, students—will be interested in doing their part in the larger crisis and would like their community to participate. They also neglect the role of the local community, which might be outraged to learn that their academic members absolve themselves from the potentially difficult decarbonization process they undertake (for more on this, please see chapter 4). Similarly, those who think campuses should not reduce their GHG contributions underestimate the role of regional, national, or even transnational forces and expectations, or else they will find themselves arguing for an awkward academic carveout from civilization-level transformations (the focus of chapter 5). My conclusion on this point is that many in academia will want to decarbonize and reduce their methane emissions, and that that motivation will add to the self-preservation goal noted previously.

We should identify and analyze who is likely to participate in such decision-making. When we speak of academic institutions making strategic decisions, we often have in mind—correctly—senior campus administrators. Presidents, deans, chancellors, vice presidents, and presidential cabinet members are the usual figures, depending on institutional culture and constitution. Private institutions also have governing bodies, such as boards or trustees, who can play decisive roles. Public institutions report to local, regional, or national governments, which can exert significant influence.

Each institution contains other populations who can enter into climate crisis decision-making. To the extent that faculty members have governance roles, professors can participate in strategy. To the extent that a given college or university responds to student voices, students can enter into the process. Apart from these formal mechanisms, faculty, staff, and students can also play informal roles through various forms of democratic participation, depending on the nature of an institution: protesting, expressing opinions through various media, listening to meetings, organizing petitions, and so forth. As the crisis unfolds, these mechanisms will change around the world as technologies develop and as institutional and public policies expand or restrict them. Beyond an academic community, other stakeholders can see themselves as having a strategic or advisory role, from scholarly publishers to churches and nearby businesses (see chapter 4). Of course, not all potential participants can participate on an equal footing. Economic class, religious affiliation, racial identity, educational attainment, geographical position, and other factors empower or weaken each potential participant's involvement and influence.

We should also anticipate strategic participation in more controversial ways. Governments may forbid some of the actions described in the preceding paragraph. Social and political movements can choose universities as sites for expression or struggle. Property damage is a likely outcome over time. Some may choose to damage or destroy property as a climate policy tactic. For example, people opposed to climate adaptation could deface or disable on-campus solar arrays. Those who see petroleum engineering as complicit in climate disasters could sabotage offices and labs. Students, staff, or faculty who view petroleum-fueled vehicles as contributing to disaster could damage or destroy such vehicles on campus. Moreover, as local conditions worsen or local perceptions of global crisis intensify, we should anticipate violence against people of all kinds. Kim Stanley Robinson's recent science fiction novel, *The Ministry for the Future*, sketches out one way this might unfold. Once climate deaths rise into the millions, a shadowy global organization conducts a range of direct actions against financial leaders at Davos, cattle ranchers, fossil fuel company leaders, air travel infrastructure, and more. We can imagine some students, faculty, and staff participating in this movement. We can also envision off-campus authorities investigating academic institutions for this very thing, accurately or otherwise. Such actions can trigger further violence through backlash or imitation. Before 2100, some campuses may experience this kind of political unrest and violence.[12]

In such a fraught decision-making context, we can also see campuses and nonacademic communities learning how to adjust their physical properties from each other. If a college or university resembles a town or small city, all can share experiences and plans as they confront related problems.

From the Ground Up

Let us start at a small level. Communities, academic and otherwise, may change how they treat artificial heating and cooling. They are likely to replace air-conditioning and refrigeration units as local or national standards and policies change. For example, adhering to the 2016 Kigali Amendment to the Montreal Protocol on Substances that Deplete the Ozone Layer may justify overhauling devices. This may lead to deemphasizing AC use, as that technology contributes to global warming through both drawing significant electrical power as well as the hydrofluorocarbons used to chill air. We could see rooms or buildings switching from AC to electric fans, which use less power and also might be more effective at cooling people than many think.[13]

These changes are not without tradeoffs and risks. Each part of an architectural system relies on the actions of many others. As Lee Gardner cautions, "Sealed windows, for example, can help a structure be more energy-efficient, as airflow can be more tightly regulated. But what happens if the electrical grid goes down in the early-September heat? A building without windows that can be opened could become uninhabitable."[14]

It is worth noting here that improvements in energy efficiency are often low-hanging fruit, as many can be realized with low-cost interventions. Insulating key pipes, switching showers to low-flow intensity, using LEDs for lighting, and other steps can significantly reduce a building's energy use.

At a more ambitious level, we may choose to create or renovate buildings based on the principle of passive cooling, which requires less in the way of powered air-conditioning. Cool roofs, which contain plants and water, can help lower interior temperatures. Green roofs in general can do this while also providing mental, culinary, and aesthetic benefits. Elmira Jamei, Hing-Wah Chau, Mehdi Seyedmahmoudian, and Alex Stojcevski describe these roofs thusly:

> Green roofs are planted with diverse plants and vegetation that are situated on top of a growth medium (substrate). The concept of green roofs was introduced to provide multifaceted benefits (social, economic, and environmental) to rapidly growing cities. A green roof often comprises several components, such as vegetation (landscape materials), substrate (growth medium), filter layer,

drainage material (for moisture retention), insulation, root barrier, and waterproofing membranes.

In addition, we may see institutions provide cooling centers on campus to offer relief to members of the community, either as standalone, separate buildings, as parts of buildings with other purposes, or as temporary structures erected for a heat wave.[15]

The areas between buildings are also subject to rethinking. Lawns may become problematic, no matter their beauty and cultural associations with undergraduate learning. Grassy areas require a significant amount of water, which is more valuable as the planet's surface heats up. Watering lawns may increasingly compete with other water demands, from local agriculture to personal bathing and drinking. Those demands can become socially visible to the off-campus world as water must be transported to those grassy quadrangles. Moreover, the use of nitrogen to sufficiently fertilize lawns also constitutes a small contribution to greenhouse gas emissions. Additionally, other purposes for lawn space may become more appealing, such as supporting trees (for cooling and for carbon sequestering) or plants (for food or insect pollination). Given these options, the institutional decision to maintain, reduce, or abolish green areas may appear as a political statement to some eyes. Institutions may not be the only actors in green spaces. Activists, both academic and otherwise, could protest or act against lawns, as we've seen with Extinction Rebellion digging up a Cambridge lawn in order to, in one participant's words, "prevent a horrible, horrible future that seems so insurmountable."[16]

Beyond campus lawns, colleges and universities may see fit to expand the number of trees on their holdings. Trees perform a fine service in drawing down and holding carbon dioxide both aboveground (in trunk and limbs) and below (in roots and soil), which has attracted many to the cause of mass tree planting to mitigate global warming. While this cause can be overstated (trees planted in high albedo areas may heat, rather than cool, them), overhyped in the short term (it takes decades for most trees to grow), or overdone (woods replacing food crops), adding trees to the earth's surface properly can be beneficial if done correctly. The idea is surely attractive to many polities and individuals, and it makes sense for many colleges and universities with suitable grounds.[17]

An additional option for those interested in drawing carbon dioxide out of the atmosphere is the emerging technology of direct air capture (DAC). This process employs machines that suck in air, sort out the carbon dioxide, then store it. For now, DAC is quite controversial. In its favor, people argue that it is an essential global strategy to reduce preexisting carbon dioxide in order to mitigate global warming's damages. Against this, some criticize the technology for being unworkable at scale without massive (and expensive) research and development plus industrial-scale deployment. Others oppose DAC because they see it as reducing popular desire to reduce individuals' own emissions. Should a campus decide to explore the technology, we could easily imagine DAC units or larger installations on campus, between or on top of buildings.[18]

Between campus green spaces today are often roads or tracks for transportation, and this opens up a major area for imagination and planning, as petroleum-fueled vehicles contribute substantially to global warming. Industries, governments, communities, and individuals already have responded to this challenge in various ways, such as successfully deploying hybrid and electric vehicles, repurposing roads for pedestrian traffic, or campaigning for people to switch from individual cars to mass transit or bicycles. We can imagine campuses worldwide following these paths to different degrees, depending on an institution's situation, leadership, and community. A university can act to reduce carbon-burning cars by offering electrical alternatives and plentiful charging stations, blocking off some driving spaces, or offering various incentives to encourage faculty, staff, and students not to use the dangerous devices. Local, regional, or national petroleum taxes might assist such moves. Some campuses might hold "parklet" events, temporarily turning parking spots into green spaces by laying down artificial grass and cordoning them off. Shifting on-campus vehicles to renewable energy sources will require a campus to support infrastructure, which may be new or expensive. We should prepare for possible increases in road damage, due to electric vehicles' heavier weight. Research shows that the leading reason drivers switch back to gas-burning vehicles from electrically powered ones is charging difficulties. Perhaps a solid campus charging capacity can help make the transition away from petroleum-burning cars succeed. Campuses can also encourage commuters to share rides and carpool. Institutional fleets can change over to hybrids or electrically powered cars, shuttles, and buses; there is already progress on this approach in primary and secondary school systems.[19]

As technologies advance over the next decades, campuses might support new transportation forms, such as cars with solar panels mounted on their roofs, allowing them to replenish their batteries while driving and to soak up electricity while parked. We can replace our asphalt roads, driveways, and parking lots with new ones that include solar cells baked into their material. A stronger policy would ban twentieth century–style cars completely, or charge them extra for entering or parking on campus. This might seem extreme in 2022, yet may become more attractive, even essential, as the Anthropocene proceeds. As we approach the year 2100, simply allowing petroleum-burning vehicles on grounds may become as abhorrent an idea as harboring mass murderers—and the former likened to the latter.[20]

Rethinking Architecture

Alongside those roads and lawns, we encounter the question of academic building renovation and construction. After all, "buildings are generally the largest user of energy and the largest source of greenhouse gas emissions on campuses," as the Sustainability Tracking, Assessment & Rating System (STARS) technical manual puts it. At present, some buildings emit carbon dioxide or draw on sources that emit carbon dioxide in the course of operations, from running heating and cooling systems to illuminating spaces and operating elevators, contributing in a tiny way to the climate problem. We can rethink campus architecture as carbon neutral, with buildings designed to not emit carbon dioxide and therefore not adding to the global load. Some architectural schools and approaches now do this, crafting buildings that on balance do not add to the climate burden in operation and whose construction process minimizes greenhouse gas emissions, including through the use of recycling. They use less power in their lives than do other buildings and draw on (or themselves contain) renewable energy sources. They may store carbon locally or off-site. They rely on materials that contain and were created with less carbon. Their insulation is stronger than what we typically built and inhabited in the twentieth century, with great abilities to block heat flow (known as "higher R-value") based on careful use of location, responsiveness to the sun, passive heating and cooling, appliances that draw less power, smart glass (electrochromic glass), and more. Some buildings may follow Rolf Disch's concept of PlusEnergy, producing more energy than they consume, either for local use in the form of a microgrid or fed back into a utility's grid, perhaps for profit.[21]

Alternatively, academia could take a different approach to building renovation, one more suited to the Anthropocene. The construction, maintenance, improvement, and eventual demolition of buildings involves not only significant waste but also carbon dioxide, most especially in concrete. We can identify amounts of carbon dioxide used in the production of building materials, in their operation over time, and held within certain items, the latter known as "embodied carbon." To reduce or eliminate that carbon would involve a great deal of redesign, changes to supply chains, inter-industrial collaboration, and more. Academic institutions might be good sites to practice, study, and even showcase such new forms of architecture. A transition phase along that path might involve more careful use of materials over time. Architect Thomas Rau calls for a "demountable" construction practice, whereby each item within a building can be identified, repurposed, and reused. Imagine a campus whose built environment consists of components that migrate between locations with buildings, or between different structures entirely, over the years. Depending on feedstock and methods of assembly, 3D-printed buildings may prove useful alternatives.[22]

Some cities and regions are considering armoring themselves against rising environmental stresses. Installing or expanding seawalls against rising waters is one such option. Some jurisdictions are elevating grounds above where they anticipate sea level rise to reach. Several cities, such as Lagos, are expanding their use of floating architecture. Another option appeared in the wake of Superstorm Sandy, when a New York design firm created Living Breakwaters, a project that includes "rocky protective structures, seeded with oysters, to reduce wave action along the eroding shore of Staten Island. These structures rebuild the three-dimensional mosaic of coastal habitat that protects us from extreme weather . . . This 'oyster-tecture' project is equal parts habitat reconstruction, climate risk reduction, and community engagement."[23]

Will campuses support and benefit from such actions or actively participate in their construction, adding their cognitive and staffing power? Alternatively, we can imagine individual universities creating their own such barriers, possibly in response to a perceived mitigation failure in the local area, or in addition to local measures. The United States Naval Academy is planning to raise a wall against the Severn River for this reason. It may also move or elevate buildings. Eckerd College in Florida is starting to plan to elevate campus buildings against a one- to two-meter sea level rise.[24]

Campus designers can draw on architectural standards as they renovate old builds and create new ones. Four different Leadership in Energy and Environmental Design (LEED) standards give detailed requirements for green construction. The STARS standards offer another set of requirements, created by academic staff. STARS addresses greenhouse gas emission in many ways, including campus inventory, sequestering, offsets, and purchasing energy from renewables. Zero Code (now in its 2.0 edition) offers another brace of detailed design specifications. The Passivhaus school of low-energy design has its own design system. Other standards may well appear over the next generation as retrofitting and new buildings face more challenging environments.[25]

In fact, we should expect new building designs to appear that will openly engage the climate crisis. As the twenty-first century advances, more ideas will emerge as human innovation grapples with the unfolding crisis. New construction materials could lower carbon and financial costs. We may revisit ancient architectural cooling designs, like the Persian and Egyptian windcatchers. There is already an emergent design school that integrates greenery and the built environment to high levels. The published plan for Amazon's new Virginia headquarters shows a vertically inclined spiral, along which ascends a rising belt of trees and smaller plants, intertwining the natural and the artificial. Milan's Bosco Verticale festoons two apartment buildings with bushes, plants, and trees. Solarpunk science fiction often offers related visions, such as residential areas intertwined with food gardens and wind turbines or large-scale solar arrays. Perhaps the best-known visual artist working in this field, the artist known as Imperial Boy portrays cityscapes covered with greenery in balcony gardens, hanging gardens, trees adorning upper floors, and roofs suffused with plants. Treetops and canopies at times overlook buildings. We can imagine new types of campus architecture following this vein. Imagine a student union with student-tended gardens on its rooftop and other levels, or faculty offices whose windows display a variety of plants. The purpose here is not merely aesthetic, but design to produce beneficial effects: cooling temperatures, sequestering carbon, adding some moisture to dry spaces, providing fresh fruits and vegetables, and offering pollination sites for insects. The contrast that design would present with, say, brutalist architecture would be impressive.[26]

The potential of universities to grow some of their own food introduces another dimension to the problem of how campuses confront the climate crisis. Globally, food production generates up to one-third of greenhouse gases, from methane produced by some animals (notably cows) to carbon burned in refrigeration, transportation, lighting, heating, and more. In response, some observers and activists call on humanity to eat less meat and fewer animal products—a request that goes against the grain of rising meat and animal product consumption worldwide. That collision between food culture and the climate crisis should play out on every campus with food service and is already starting to. Recall that many academic institutions serve meat, dairy, and general animal products in dining halls and for meetings and events through some mixture of outsourcing and in-house provisions. Some campuses also raise animals for these purposes as part of agriculture and food system programs. All of this contributes (in a small way, within the global context) to climate change.[27]

If campus leaders choose to, colleges and universities can encourage their populations to eat a vegetarian or vegan diet through incentives and changing what menus provide—that is, offering fewer animal-based foods and more plant-based ones. Britain's Tate Modern museum took steps along these lines, announcing that "the museum's restaurants will prioritize sustainable food-sourcing strategies and will offer more vegetarian and vegan meal options." Institutions can emphasize locally produced food in order to reduce travel and storage costs, as those involve carbon emissions. Colleges can serve artificial meats like the Impossible Burger to help transition diners to new diets. Food providers could add charges to items, either directly (at point of sale) or indirectly, based on the calculated carbon impact of individual food items. Taxes on some or all meat may appear in such charges. To some extent, administrators will not be leading their populations but responding to their communities' changing tastes as eating habits change. Students, faculty, staff, and campus visitors may alter their diets in response to the crisis and request suitable provisions on campus.[28]

I don't want to give the impression of a smooth, speedy, and total switch to veganism throughout global academia. Obviously, there are many obstacles to such a transition. Psychologically, we view food as radically part of our individual selves; culturally, we often locate our food habits within various ethnic, religious, geographical, or other traditions. Think of the cultural politics around societies that value hunting to provide meat, or the many ways political clashes start because of or use food, like opponents to the Green New Deal charging it with taking steaks away from would-be eaters. Pushing against those deeply set gustatory forces almost universally guarantees friction. Moreover, beyond academic administrators making food service decisions, campus activists may also call for the same through protests, demands, sit-ins, or other means. Some may take to social media to shame meat-eaters and celebrate their opposite numbers.[29]

Sign in a residence hall window, Georgetown University, 2020. Photograph by the author

Other campus services may require redesign in the face of climate pressures. Consider, for example, the carbon footprint of university health services. They consume a great deal of petroleum-based plastics, often in single-use ways. Some of their technologies require significant amounts of electrical power, such as advancing sensor gear. Does a campus decide that the human benefits are ethically worth the GHG costs, or does it seek to reduce the carbon emissions its health care service is responsible for?[30]

Let us turn to buildings beyond clinics, cafeterias, and dining halls. How to power those buildings is a key question higher education must confront. As of the writing of this book, the supermajority of colleges and universities source electrical power off-campus to utilities or companies through power grids or other connections. Those energy providers often rely on carbon emissions to work, burning coal or oil. Not all do, of course, as renewables (wind, solar, hydrological, waves, geothermal) produce a small but rising amount of electrical power. Organizations and communities of all kinds increasingly face moral and practical questions over their reliance on carbon-emitting power providers, and we should not expect academic institutions to be exempt from such scrutiny. To begin with, how will campuses obtain power off-site? How many will have the option to simply stop sourcing from coal, oil, and gas? Available sources are growing, but are by no means plentiful around the world, nor are they necessarily economically affordable.

This brings to mind a second option: for postsecondary institutions to start or to increase providing their own electrical power. The usual practice is not to do this, with the exception of the University of Texas at Austin's self-hosted power plant, yet the crisis can change this habit. Some colleges and universities have already installed solar panels, either on top of buildings or as free-standing arrays; it is plausible to imagine much more of this, especially as installation prices drop and public interest rises. This is, of course, more appealing on campuses that receive more sun yearlong. Similarly, campus grounds may feature wind turbines—again, depending on the physical affordances of a given area. Hydropower may also work for those institutions with access to flowing waters, from rivers to oceans. We can imagine various ways of supporting such technologies, as with student workers who learn potential career skills.

Ultimately, we can envision institutions completely banning carbon burning of all kinds on campus. This would be a detailed and difficult achievement to accomplish, given the many ways we've seen carbon emissions appear on campus this chapter. To give a sense of how much can be involved, in 2020, one American town considered a law that prohibited carbon burning within its zoning limits and that would "ban all gas appliances, close fossil fuel pipelines, and move gas stations outside city limits":

> Starting from 2030, all water heaters, space heaters and stoves that rely on gas would have to be replaced with alternatives. Fossil-fuel-based leaf blowers would be phased out with incentive programs and eventually outlawed, while gas stations would be asked to convert to electric charging or relocate. The proposal promotes electric vehicle use but does not include a ban on vehicles that run on fossil fuels. In the shorter term, the city would mandate that all buildings, including single-family homes, meet specific energy requirements by 2029 and upgrade all lighting to LED by 2022.[31]

On the one hand, such a proposal looks daunting or, to some, foolhardy. Yet the rising pressures of the climate crisis may shift our perceptions so that a campus ban looks interesting, practical, or necessary. Eventually, not having implemented one may look reckless or even criminal.

Running on top of an academic institution's electrical power infrastructure are its computing systems, which presents all institutions with a strategic choice. In one future, universities expand their use of information technology as the climate crisis expands. Reduced travel in order to avoid emitting more carbon yields more remote work, as we saw during the COVID-19 pandemic, which means more computing across the board: more data being stored and accessed, more processor use, more graphics (for video and virtual reality), more throughput overall. Additionally, some climate research requires massive computational power for analysis and simulations involving large amounts of data and complexity. We can use more networked sensors and devices within the Internet of Things, a still-developing project to connect more and more objects to the internet while expanding their ability to gather data. Such a vast computing stratum can, among other things, obtain more and better data about our greenhouse gas emissions and environmental impact.[32]

In another future, campuses, and possibly society, use less information technology. Some recent research has found that certain types of computer processing, such as training neural networks on large datasets for long periods of time for artificial intelligence purposes, use dangerously high amounts of electricity. Blockchain technologies, especially bitcoin, use large amounts of power, particularly in mining new bitcoins. Cloud computing facilities, while increasingly efficient in their power use, nonetheless continue to grow in size and electrical draw. One study examining "three major environmental footprints (i.e., carbon, water, and land footprints)" of cyberinfrastructure determined that

> data storage and transmission emit . . . 97 million [metric tons] of CO_2e a year—roughly equivalent to the annual carbon footprint of Sweden and Finland combined. Similarly, the median global water footprint of Internet use is estimated to be 2.6 trillion L of water, or the equivalent of filling over 1 million Olympic-size swimming pools. Finally, the median land footprint of Internet

use is approximately 3400 square kilometers of land, representing the combined size of Mexico City, Rio de Janeiro, and New York City.[33]

In short, the very technology we hope to use more of in order to reduce carbon emissions may actually emit enough carbon dioxide to be globally dangerous. If we perceive that to be the case, we may choose to reduce some or all of our computing practices, reducing a century-long trend of ever-expanding IT usage. We could choose to create and consume less-demanding forms of digital content: smaller documents, smaller websites, less animation and video, more text. Our hardware replacement cycles—a new laptop every few years, a new iPhone annually—might attenuate.[34]

This might take the form of green computing. This way of building, operating, and thinking of digital technology might value efficient power use in production and assessment, as well as greater inclusivity. Green computing includes assessing and mitigating environmental impact. As the Massachusetts Green High Performance Computing Center (MGHPCC) describes its work: "Environmental design for LEED Certification requires attention to numerous details, including construction methods and materials, landscape and site design, and water conservation. For example, 97% of the construction waste generated while building the MGHPCC was recycled or reused instead of going to landfills; materials high in recycled content were used wherever possible, and landscaping was designed to minimize water use and storm water runoff."[35] Note the point about building data centers to high standards of efficiency, such as LEED. Connecting architectural standards to green computing is just one example of connecting different climate change responses.

Choosing such a future might not be an act entirely under our control, if climate problems impact our infrastructure directly. Rising waters, storms, and fires could destroy cables and buildings. Crucial rare earth elements may become more expensive as they become scarcer. Carbon taxes could make computing usage more expensive, which might drive a multitiered digital experience based on economic class. Similarly, internet speeds might slow, except for those who can afford the best. In short, the green computing choice might be made for us. Campus computing may be both lighter and more green, perhaps not so powerful as it is in 2023.

Similar questions may confront academic communities when it comes to financing these physical transformations. Divest from fossil fuels—and then? Shift scholarships to student populations most at risk of climate disaster?[36]

It is worth repeating that any redesign to a physical campus runs the risk of opposition both from within and beyond its community. Altering buildings, even for the great cause of the climate crisis, can run afoul of historical designations, especially to the degree that those structures are subjected to detailed codes and regulations. Other forms of cultural opposition could arise from a community's sense of sentimental value, or as a political statement in opposition to whichever coalition enabled architectural transformation.

Opposition may also meet another use of academic buildings beyond their traditional teaching and research roles. Beyond supporting members of a college or university community, the academic built environment may also house people for dire emergencies. Emergency workers and first responders can stage from a campus, as they did in 2019 at Australia's Charles Sturt University when fighting fires nearby. Civilians may flee to the safety of a college or university. (The reverse can also occur, as colleges and universities evacuate some or all of their population to safer ground, as occurred with Sonoma State University during the 2019 fires on America's West Coast.) Further, campuses may become sites to house climate refugees as disasters and chronic stresses drive people to migration. Academic buildings can be repurposed, while residential buildings are obviously well suited as housing. Campus leadership or its community might urge making these sites available. Community or political pressures might exert themselves along the same lines. Forward-thinking institutions could plan ahead to identify spaces for climate refugee hosting, even adding this purpose into renovation and construction plans. At the same time, such institutions should anticipate opposition to such moves from a variety of sources. On campus, such resource allocations can compete with academic community desires. Moreover, early twenty-first century history has shown many times that societies are quite capable of meeting migrants with opposition and outright hostility; both on-campus and nearby populations can express those views.[37]

Migration might not be limited to individuals and populations. As the climate crisis worsens, entire academic institutions may consider relocating partially or entirely to safer terrain. Aridification, desertification, rising waters, rising wet-bulb temperatures, droughts, and decaying local economies could all compel a college or university to find new grounds. Obviously, these aren't snap decisions to make with immediate results. Long-term planning is the context for such strategy. Individual institutions will have to assess their risks and determine better sites. They will partner with local communities, both their present and future hosts. To the extent the crisis worsens, we could see campus after campus relocate. We might see a widespread academic migration occur as the rest of the twenty-first century unfolds: universities in flight.

CHAPTER 2

Doing Research in the Anthropocene

Our knowledge of the Earth's system is about where our knowledge of the human system was at the turn of the [last] century. We have so much to learn—but so little time to learn it.

—JAMES LOVELOCK

In the previous chapter, we explored how the climate crisis may impact the physical campus. Now, we turn to one of higher education's essential purposes and how that connects with climate change: the scholarly research mission.

Not all colleges and universities conduct research to a similar degree. Some institutions make research their leading purpose, such as the sort labeled "research intensive" or "Research I universities" in the Carnegie Classification system. On the other extreme are colleges and universities that choose teaching to be their main function and deemphasize faculty scholarship. In between are a range of institutional configurations. Moreover, within a campus, we can see wildly differing levels of research commitment. Indeed, within a given department there may be a broad spectrum of scholarly inquiry and output.

Overall, though, academia as a whole conducts research and publishes results in a major way, advancing human knowledge and enhancing our collective life, and this is where the climate crisis comes in. Academic research plays a crucial part in our understanding of the Anthropocene at multiple levels. It gives us the best projections of what is likely to occur next and of our options. Academic research and development also add to our options, yielding tools and practices we can apply to address the warming world. Scholarship may, in fact, represent higher education's greatest contribution to how humanity grapples with the climate crisis, and it therefore deserves our full attention here.

To begin with, we should outline how the structure of academic research maps onto the crisis. Our enterprise is divided into disciplines and subdisciplines, with fields and niches within them. Individual branches of knowledge address the climate crisis through their particular intellectual frameworks, from the natural sciences to the social sciences, humanities, and the arts. Many have already begun work, as we will see below. While many researchers work separately, plowing disciplinary furrows, they do at times connect in an interdisciplinary way. As we will see, climate change as a topic appears to have a strong ability to evade easy silo-ization. Collaboration across boundaries is a path of its own. Additionally, we see criticisms cross those furrows, as when researchers in one field critique the work of others—for example, ethicists calling into question a geoengineering project.

As we progress further into this Anthropocenic century, each preexisting academic discipline may change as it grapples with climate change. Each academic unit's research agenda can develop new themes and topics. The pressure to do so may come from within as individual scholars, then groups thereof, determine new courses of study. If we consider students to be active members of a discipline, then they can exert pressure by their demands for classes as well as their own research interests. External forces can also come to bear, sometimes in a demand for more work along certain lines as governments, funders, alumni, or the immediate community call for expanded or new inquiry.

At the same time, we should also expect demands for reductions in research to be expressed. To pick perhaps the most obvious example, activist organizations could call for the curtailment of petroleum engineering as a field, seeing that field as devoted to the further exploitation of buried carbon resources and hence an accelerator of the climate crisis.[1] On-campus student movements may pursue a similar agenda, as may fellow faculty members, staff, politicians, local authorities, or various people within the immediate community. Other fields deemed too close to carbon could also draw outside scrutiny (see chapters 4 and 5).

Another course of research reduction may follow the logic of opportunity cost, with community members, activists, staff, state officials, faculty, students, and so on protesting faculty who research something other than climate change. Scholars examining medieval French literature or the qualities of newly observed exoplanets may see their research agendas' reputation shift from being perceived as necessary components of a university's wide-ranging research to being deemed wastes of precious resources (time, money, human thought) during a global emergency.

We can start anticipating these alterations to the research enterprise by exploring how individual fields are approaching, or starting to consider, climate change. We may already see the first traces of an emergent, broad-ranging, multidisciplinary research agenda for colleges and universities. An examination of current publications reveals the breadth of climate research as well as its growing interdisciplinarity.

One caveat: what follows is by no means exhaustive in terms of research examples. There are many, many papers, monographs, posters, datasets, videos, podcasts, simulations, and presentations to cite. This chapter instead outlines research directions in disciplinary housings at a top level, with sample work only to support each point.

The Natural Sciences

In general, these are the fields most people are likely to think of as key to researching climate change.

Climate science should certainly grow, of course, as the academic branch (or cluster of academic branches) leading the way in our collective understanding of climate change. The field will continue to develop, adding more data and models. Climate science conducts extensive field research worldwide to gather evidence from ice cores to sea temperature and atmospheric humidity readings. It also relies on advanced computational power for simulations of current and historical processes as well as crucial projections into the future. Similarly, and obviously, earth science and environmental studies are equally central to climate investigation, since the environment is precisely what is being transformed. A few examples can suffice for our discussion for now: determining how rising temperatures will reshape electrical power grids, examining the carbon sequestration capacity of the Amazon basin, projecting flood damage from melting Himalayan glaciers, and decreasing groundwater feeding wells. New branches of these fields are already appearing, like environmental assessment and paleoclimatology. We should expect to see more of this kind of disciplinary hybridization. Altogether, these fields detail and adumbrate the full extent of climate change.[2]

Other scientific fields bring powerful intellectual firepower to bear on the subject. One related and clearly vital field is meteorology, given the vital importance of understanding and forecasting storms, the nature of which is changing as the global system heats up. This includes, for example, attempting to rapidly determine to what extent climate change drives current storms. At a broader level, this field studies the planetary flows of weather as they change in the Anthropocene. Oceanography delves into the immense seas, one of the most powerful components of the earth's systems, from serving as heat and carbon sinks to changing temperature and chemical compositions at different depths, connections between rivers and polar ice caps, how the Arctic warms more quickly than the other seas, the interaction of oceans with atmospheric humidity, connections between tides and polar methane emissions, and the vital Atlantic Meridional Overturning Circulation (AMOC). Other fields focused on water, such as hydrology or lake studies, shed light on changes to glaciers and how bodies of water hold or emit carbon, how salt water can infiltrate freshwater as sea levels rise, and how to rethink carbon budgets.[3]

Still more sciences play major roles in studying the impact of climate change on the world. Geologists examine the movement of carbon through different materials within the Earth's crust, yielding insights about how carbon dioxide might be retained or loosed into the atmosphere. Geographers conduct vital work into how the Earth's surface changes, such as mapping and projecting the Antarctic ice cap's instability, understanding changes to land ice, and modeling likely results of humanity failing to stick to the Paris Accords. Astronomy and astrophysics add a range of space-centered information, such as analyzing the degree to which Earth's albedo has changed in recent years and how that suggests more global warming to come.[4]

The life and earth sciences play key roles in investigating global warming, unsurprisingly. Biology is already examining the appearance of new or changed animal diseases, driven by climate alterations. Similarly, the loss of biodiversity caused by climate shifts is a topic for which biological science is well positioned. Branches of biology focus on particular strands of the web of life, exploring how climate change impacts them. For example, one study analyzed how global temperature increases cut down one seabird population by pressuring closely related fish and plankton species. A recent set of studies analyzed a potentially catastrophic decline in insect populations and found climate change to be one cause: "Climate change, habitat loss and degradation (especially of tropical forests), and agriculture emerged as the three most important stressors considered by our authors, with the first of these receiving the greatest attention in the symposium."[5]

Agriculture, food science, and forestry play related research roles. Food systems constitute an essential element in causing climate change, as certain elements (raising cows most notably, as well as emissions from nitrogen fixing) contribute methane to the atmosphere. Altering those systems may be equally essential in mitigating disaster, from changing fertilizer composition and human diets to expanding composting practices. One mitigation strategy involves repairing and rebuilding billions of acres of soil worldwide, which involves not only agriculture but also forestry, as trees capture and sequester carbon dioxide. This forest capacity appeals to many, since it is very low cost, low tech, and also visually pleasing; understanding its functions and projecting its possible implementations is the work of forestry and allied fields. Multiple research efforts research and develop tools along this line, such as creating an index of forest quality, against which one can plan supportive work, longitudinally mapping global forest data, assessing what kinds of trees store the most carbon, or measuring to what extent global forests serve as net carbon sinks.[6]

Elsewhere in the sciences, engineering stands out for its many capacities to work within climate change and climate mitigation. The civil engineering branch already offers techniques and devices for defending areas against natural threats; as those threats increase, civil engineering should grow in prominence. Mechanical engineering comes to bear as humanity develops non-carbon-based energy generation, transmission, and storage mechanisms, from wind to solar, hydro, nuclear, biomass, and more. We can see that branch of engineering at work in the effort to develop carbon-capture technologies. One Purdue University team developed the brightest white paint ever created by humanity, which might prove useful in increasing structures' reflectivity and hence decreasing heat absorption. Certain parts of engineering should grow in size and importance as the climate crisis grows. For example, coastal engineering is crucial to helping coastal communities deal with rising waters. Mechanical engineers develop new ways to reduce human carbon output, such as applying specific paints to cool buildings or analyzing the carbon cost of large-scale indoor cannabis growing. Academic institutions can play active, hands-on roles in supporting engineering's approach to climate change. One example comes from a group of American universities that for years has hosted a testbed for innovative and sometimes enormous wind turbines. Of course, environmental engineering stands out in this context, as with a multinational team that compared two different ways of providing power to Southeast Asia, hydrological (dams) and solar.[7]

Transportation engineering has come into focus due to the role of petroleum-based vehicles in pushing up the global CO_2 count. That field is now engaged in addressing this problem. Aviation, naval, and ground transportation researchers and developers are rethinking materials, trying out new ways to design vehicles that burn less carbon. They also examine what enables drivers to choose gas-powered, hybrid, or electric vehicles. At the same time, transportation infrastructure faces challenges due to rising temperatures, such as rail lines and roads warped by melting ice or shifting soil. Engineers in these fields work to determine exactly what is occurring while offering new ways to cope with the crisis. Some transportation studies have argued for the necessity not just of ending our reliance on cars that burn fossil fuels but of getting past cars entirely.

Computer science adds to the innovation mix, especially as it plays a crucial role in the development of climate science. Advanced software, sensors, networks, applications, models, and visualization are key tools for researchers measuring climatological changes and trying to extrapolate forward from them. Large amounts of data and the complexity of climate dynamics require advanced computation. The computer science field has already responded, creating new tools and approaches, such as using machine learning or creating vast simulations to model planetary system dynamics. This branch of the sciences faces a particular challenge in the Anthropocene based on its potential contributions to either mitigating or exacerbating global warming. On the one hand, there are now many options for relying on the digital world to decrease carbon use, as we've seen during COVID-19 quarantines (for example, using video recordings and conversation instead of flying). Cloud data centers may be more efficient in handling the electrical costs of data than local computing clusters. On the other hand, there is also a rising argument that digital services draw too heavily on carbon themselves. Advanced AI projects, which require good numbers of processors working at high speed, may be criticized for worsening, rather than ameliorating, climate change, as noted in chapter 1. For example, training large programs to simulate human conversation involves hefty amounts of computational power, which may either draw from carbon-burning sources and hence contribute to global warming, or rely on alternative energy sources that might be better used for more socially beneficial purposes. Some researchers claim that graphics- and audio-intensive videoconferencing tools are also dangerous drains on resources. Creating blockchain content may also be similarly critiqued, especially with regards to mining bitcoins. Some argue that we urgently need innovations in materials sciences and hardware construction in order to reduce the carbon drain caused by computation, especially by growing AI applications. Others have called for related low-carbon approaches to computers and computing, arguing for redesigning computer gaming to reduce carbon usage, for reconfiguring artificial intelligence to emphasize efficient power use ("green AI"), or for green information technology in general. Working toward such goals can involve developing new ways to measure computation's use of energy and its impact on the world.[8]

The Social Sciences

Climate change is not a problem of nature but more that of people.

—ALEXANDER FEHER, MARTIN HAUPTVOGL, PETRA TANGOSOVA, AND LUCIA SVETLANSKA[9]

The natural sciences are not the only division of academia that has a powerful research interest in climate change. Many branches of the social sciences do as well, which only makes sense. What other fields are so committed to studying human behavior in its full range of responses to the Anthropocene?

Perhaps the most visible instance of this idea is in the field of economics. Economics is already investigating many economic dimensions of the Anthropocene as climate change changes economic externalities, how economies function, and how we think about them. Yale University professor William Dawbney Nordhaus won the 2018 Nobel Memorial Prize in Economic Sciences for his work in modeling how humans discount the future as compared to the present, setting up a way for us to understand the actions we do or do not take in response to the chronic climate crisis. Nordhaus has come under criticism for this work from multiple quarters, with critics charging that these discount models make it easier for humans to plan only within very short-term horizons. In another example, one multinational research team examined China's progress toward restraining its carbon output at levels that would prevent a temperature rise of about 1.5 degrees Celsius, while another group interviewed policy leaders worldwide to determine the impact of the COVID-19 pandemic on attitudes toward decarbonization.[10]

Other branches of economics may expand or alter preexisting research strategies. Developmental economics could become something quite different in the light of climate change tensions along the global North-South dimension. What will developmental economists recommend for national trajectories as the globe turns away from burning coal? How will that field account for an international aspect to climate justice? Elsewhere within the house of academic economics, finance is already exploring climate change angles, including creating new hedge strategies. One research team developed models for understanding how financial institutions may respond to various forms of macroprudential regulation. Carbon pricing has long been one way to incentivize humans to use fewer processes that emit carbon dioxide; many studies have proposed or probed precise mechanisms for making this method effective.[11]

At a different level, economics is grappling with new, large-scale models for understanding the global economy. There is growing discussion in European and Asian academia about the circular economy idea, which reduces our current emphasis on perpetual growth and emphasizes instead greater reuse and recycling of present assets. A related concept is Kate Raworth's model of a doughnut economy, tasked not with overall growth but with targeting economic resources for the neediest parts of a national or regional economy. At the same time, the doughnut economy reduces economic activity in domains that threaten ecological balance, including but not restricted to climate change. Other economists call not only for ceasing economic growth but for reversing it as part of a degrowth agenda. The goal is to seriously cut back civilization-level carbon production. These discussions overlap with, and help fuel, explorations of new metrics beyond gross domestic product for measuring economic health.[12]

The legal world within universities may respond to global warming in a similar fashion as economics seems to be doing. After all, it is not surprising that laws and legal practices might change as the world does. The historical record demonstrates this. We have already seen legal explorations along these lines. For example, who is liable for legal challenges as severe weather events escalate? Do we need new categories of legal action to account for the unusual structures of climate change events? Can the practice of giving nature a legal voice become widespread? Some lawyers and activists have called for "ecocide" to enter legal codes. Legal scholars may need to confront such emerging ideas as they frame out the intellectual course of law for the next century.[13]

If law and economics represent social sciences grappling with climate change through research, we should expect political science to also participate. Political scientists have indeed already begun to do so across a range of that discipline's scholarly registers, including comparative politics. Some have extended the ancient political concept of sovereignty to project different ways national governments may cooperate, combine, or work in isolation to address (or refuse) the crisis. One possible way forward is for the international community to build a massive new global superstate in order to organize humanity's response to climate catastrophes. Political scientists are well positioned to think through this vast problem. In the opposite political direction is bottom-up change moving through a polity; political science can analyze how different decision-makers within civil society make climate choices without top-down state control or direction. International relations provides a vital lens through which to see all kinds of climate stresses and opportunities, such as coalition building, border-crossing climate refugees, international trade in crucial materials, and bilateral negotiations for shared post-carbon energy development. To pick one example from the climate justice world, should an international agreement settle reparations on the nations that contributed little to, but are suffering the most from, global warming? We are already seeing new dynamics in geopolitics as the climate alters the playing field. For example, an ice-free Arctic opens up possibilities for shipping and seafloor mining, which in turn entail multilateral negotiations and power-jockeying across the North Pole. The Antarctic, while obviously not ice free, is famously losing ice and becoming more valuable as a potential resource center and therefore zone of geopolitical competition. Academic understanding of these radical changes could be crucial for policymakers and citizens.[14]

Climate activism is a ready subject to document and analyze, from people working within mainstream political parties to those on the fringe, from lobbying and mass movements to hacktivism and direct action. Political science has much to offer our consideration of groups like 350.org and the Sunrise Movement. The concept of people seeing themselves as "climate citizens" stems from political science, as does its development. Thinking futuristically, political science can help us anticipate—or even help build—an emergent politics whose contours are still fuzzy at present. Leah Cardamore Stokes notes Greta Thunberg recommending that we "try to push for a political movement that doesn't exist. Because the politics needed to fix this doesn't exist today." Political science can help us think such politics into reality.

Other social sciences are also gearing up to research climate change's impact on human beings. Some psychologists are studying mindfulness and "climate grief" as a way of understanding how we cope with environmental transformation. Psychology can draw on previous work on the mental damages caused by natural disasters. As the American Psychological Association warns, "Climate change–induced disasters have a high potential for immediate and severe psychological trauma from personal injury, injury or death of a loved one, damage to or loss of personal property (e.g., home) and pets, and disruption in or loss of livelihood."[15] Other psychologists see that fearsome environmental change eliciting mental echoes even in advance of truly destructive changes. They describe a kind of anticipatory dread and offer neologisms to capture that sense, such as planetary dysphoria or "solastalgia":

> While dispossession and forced separation from home are potential triggers for environmentally induced distress, what about similar distress in people who are not displaced? People who are still in their home environs can also experience place-based distress in the face of the lived experience of profound environmental change. The people of concern are still "at home," but experience a "homesickness" similar to that caused by nostalgia. What these people lack is solace or

comfort derived from their present relationship to home, and so, a new form of psychoterratic illness needs to be defined.

The word "solace" relates to both psychological and physical contexts. One meaning refers to the comfort one is given in difficult times (consolation), while another refers to that which gives comfort or strength to a person. A person or a landscape might give solace, strength or support to other people. Special environments might provide solace in ways that other places cannot. Therefore, solastalgia refers to the pain or distress caused by the loss of, or inability to derive, solace connected to the negatively perceived state of one's home environment. Solastalgia exists when there is the lived experience of the physical desolation of home.

Ecological grief and ecological anxiety may have the greatest impact on Indigenous and otherwise marginalized people, being all too often at climate change's cutting edge. Several have called for a "new field . . . ecopsychology" and labeled one set of behaviors "ecoanxiety." Others have coined related terms like "geotraumatics." Lise Van Susteren argues that anticipation of climate catastrophe can be so damaging to the human mind as to constitute *pre-*, as opposed to post-, traumatic disorder.[16]

Sociology and ethnic/race studies scholars are examining how different social groups perceive and are impacted by the climate threat. The history and present politics of marginalization have given rise to calls for us to think of the crisis in terms of "carbon inequality." One recent study, for example, argues that Black and Latino people—often disproportionally represented in imprisoned populations—are more likely than the rest of the population to suffer adverse reactions. Demographics offers a vital perspective on how human reproduction and mortality may change.[17]

Seen within the social sciences, the field of education is already researching how to teach climate change, as well as how the climate crisis may impact schooling. Obviously, this is the topic of the present volume, and pedagogical concerns are treated in the following chapter, but here we should signal the particular research interest within the education field. One key topic to research: what are we currently teaching, and how is that working to date. Education faculty can also analyze how academia responded to COVID-19, using that experience to generate a glimpse of how colleges and universities might react to the climate crisis. We have also seen research into the impact of institutional endowment divestment campaigns and campus planning mechanisms for climate disasters.[18]

The Arts and Humanities

We began this chapter with the natural sciences, which make intuitive sense as the leading disciplines for climate research. We expanded our survey to include the social sciences as fields examining the many ways humanity does and may respond to global warming. This brings us to the arts and humanities and the question of what research they can contribute to the climate crisis.

Literature scholars can track and analyze emerging climate change texts, such as those emerging from the solarpunk subgenre within the science fiction genre, mainstream climate fiction ("cli-fi"), or games that simulate and teach climate topics. Afrofuturism may appear as one such emergent textual corpus, a visionary way to combine racial and social justice with climate change. On a broader front, literary researchers can also study the appearance of climate across texts that aren't nominally concerned with the subject. Indeed, we can imagine many ways a kind of climate-inflected literary historicism emerges. Closely related may be linguistics, which could scrutinize the ways languages describe climate change, as well as the larger possibility that languages can change under the multileveled impacts of climate change.[19]

Along similar lines, some parts of the field of communication are already working on a key goal: how to best share information about the climate crisis with various audiences. There are already faculty members who have worked in this subdiscipline for years. Yale and George Mason Universities both have established climate communication centers that conduct this kind of work. For example, one recent project creates a geographical model showing which urban areas today are most like what a given site will be after two generations of climate transformation: "In the eastern U.S., nearly all urban areas, including Boston, New York, and Philadelphia, will become most similar to contemporary climates located hundreds of kilometers to the south and southwest. Climates of most urban areas in the central and western U.S. will become most similar to contemporary climates found to the south or southeast."

On a similarly visual note, David Wallace-Wells explores the difficulty in communicating remote disasters through photographs.[20]

On an institutional level, research into more effective climate communication may also be a way to build support for academic scholarship internally and from outside sources. This will likely be especially salient for researchers running into political opposition and financial stresses. As Kathleen Fitzpatrick argues, a public that has to some extent lost faith in higher education should be persuaded that supporting the academic enterprise is worthy, a process that will take a great deal of care and time.[21]

Another field focused on communication, journalism, is also already engaged in exploring how to express the climate crisis through its particular media. A University of Colorado project tracks how major newspapers cover changes to the polar ice caps. Two European researchers analyze the role of temperature changes in news stories about the climate. Emily Atkin creates digital visualizations of climate impacts to reach a mass audience, with multimedia documents like one concerning the Bolivar Roads Gate System. Others pose a journalism ethics question: to what extent should news content play an advocacy role, rather than reporting multiple sides of a conflict? Meanwhile, journalists like Christian Parenti with academic positions tell the climate crisis story as it occurs, crafting the Anthropocene's first draft. We can anticipate student journalists questioning institutional data, as University of Florida undergraduates did with that campus's published COVID-19 data in 2021. One Nieman commentator calls for climate change reporting to become central to all journalism in a piece provocatively titled "If You're Not a Climate Reporter Yet, You Will Be":

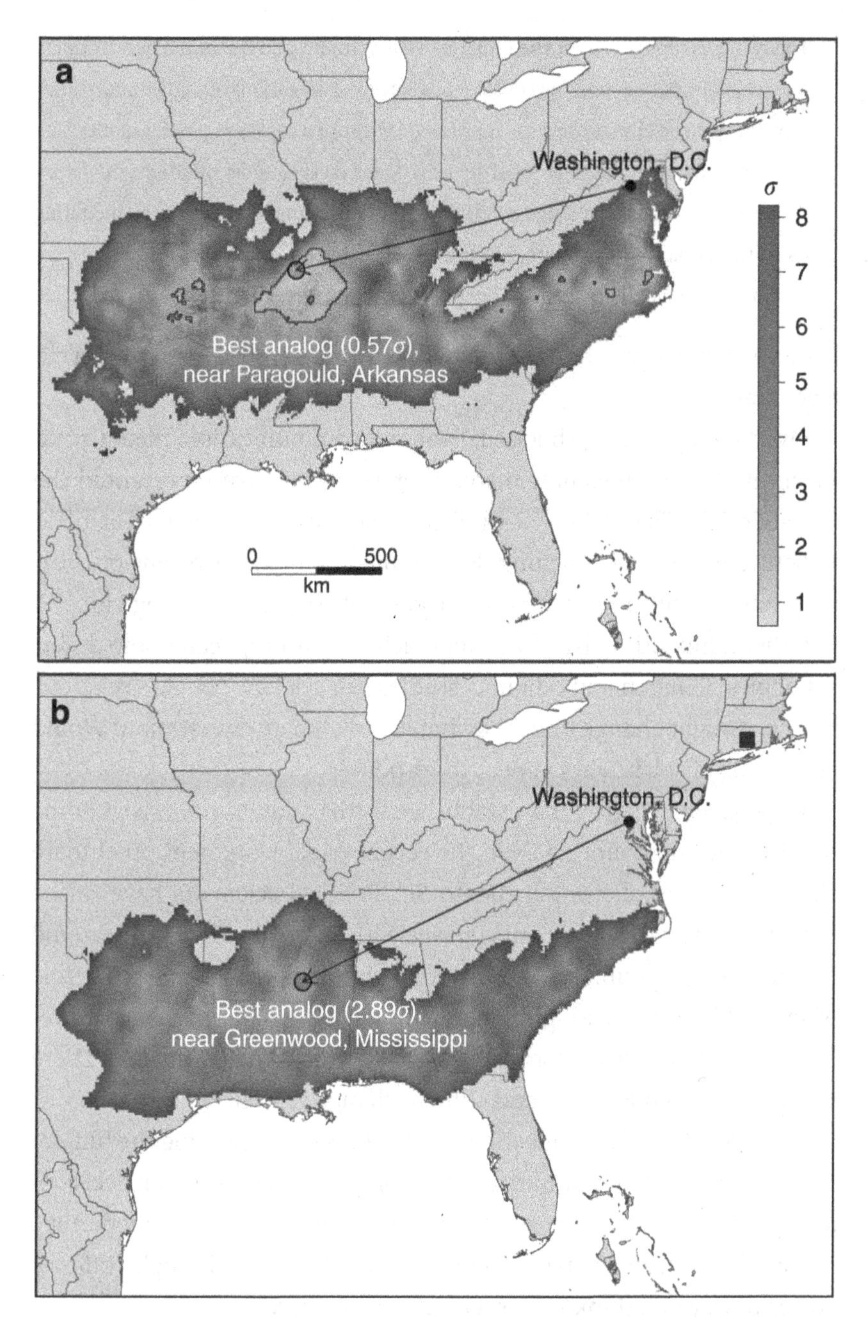

Visualizing how climate change may feel over time by way of spatial comparison. Matt Fitzpatrick, University of Maryland Center for Environmental Science

> Today, news reports about the climate crisis primarily come from a newsroom's science, politics or economics desk. A few news organizations already understand, though, that the climate crisis is more than a beat or a topic—it poses urgent questions that affect all sectors of society. Based on this understanding, journalistic coverage of climate change needs to involve all teams of a newsroom, including its culture, finance, real estate, lifestyle, fashion, health, and sports journalists . . .
>
> Climate journalism needs to become just as integrated in every vertical.[22]

Note the newsroom echo of academia's growing interdisciplinarity on climate matters.

Overlapping with both journalism and communication, media studies similarly examines how media represent and have represented the climate crisis. That field also studies connections between media infrastructure and global warming. For example, a series of recent publications has sought to quantify the carbon footprint of streaming video.[23]

Other humanities disciplines approach the Anthropocene from different angles. Religion and religious studies can ask how people's religious behavior might change under the impact of climate devastation. Would new forms of spirituality appear in response?

How have or might long-established faiths address the crisis? Some scholars will no doubt analyze the recent papal statement on climate change and Catholicism, "Laudato Si." Political scientists have raised the topic of how religions could engage with political movements around climate change. Similarly, one Turkish philosopher wonders what role Islam could play in creating an organic, non-colonial climate change movement. Some research explores a Taoist approach to climate change. There is research into Buddhism and climate change as well.[24]

Philosophy as a discipline offers many ways of considering the unfolding planetary emergency, starting with applying ethics to human decision-making. A 2010 collection probed climate change in terms of modernity, human finitude, subjectivity, technology, and ethics. Philosophy offers a powerful way into thinking through climate justice.

Philosophers can also turn their gaze on other disciplines, examining how they think through the crisis.[25]

Women's studies is energetically involved in thinking through climate change—appropriately, since it was a woman scientist, Eunice Foote, who first identified the mechanism of global warming in 1856. This field draws attention to the different and often unequal ways women and men experience and respond to the climate crisis. A women's studies perspective also analyzes the all too unequal role of women within climate science and activism.[26]

In addition to the humanities, the academic arts also play a role in researching climate change.

Photography can effectively communicate the details, scale, and impacts of the Anthropocene, even using different photographic techniques and technologies to reveal information that often escapes the human eye. Installation artists can encourage audiences to rethink how they approach the global crisis, such as getting us to rethink anthropocentrism, as Anicka Yi has done, or to provocatively imagine a home after climate change's ravages, as a Superflux installation did at the ArtScience Museum in Singapore. Creative arts projects can help us imagine how the climate crisis might play out, and what the human response could be. For example, the solarpunk concept allows artists to visualize more positive Anthropocenes, including urban areas rich with green life. Belgian designer Luc Schuiten coined the term "archiborescence" to name his visions of built environments deeply blended with plants and trees. Heidi Quante and Alicia Escott's Bureau of Linguistical Reality

> is a public participatory artwork . . . focused on creating new language as an innovative way to better understand our rapidly changing world due to man-made climate change and other Anthropocenic events. The vision of the artwork is to provide new words to express what people are feeling and experiencing as our world changes as climate change accelerates. We will be using these new words to facilitate conversations about the greater experiences these words are seeking to express with the view to facilitate a greater cultural shift around climate change.

The bureau has hosted a series of events, including many academic participants, and generated provocative, futures-oriented terms like pregret, premorse, apex guilt, Myopecene, Slowpocalypse, and Phantom Species Syndrome.[27]

Museum studies and preservation encounter this topic from another perspective, as they grapple with how to preserve potentially endangered objects and sites from environmental hazards.

How can we protect and maintain historically and artistically vital spaces threatened by rising seas, desertification, or excessive heat? Practically, museum studies and anyone engaged in art preservation have to contend with using technologies and techniques, as well as developing new ones. On a conceptual level, this academic branch faces deep questions about authenticity and reproduction, the importance of setting, and the viability of virtual experience. Perhaps there is a conflict between a Romantic or Beat movement conception of authenticity versus postmodern practices like DJing, remixing, fast fashion, and the decontextualization digital media brings about. Are architectural casts and plaster reproductions worth returning to in order to render broader access to scarce originals? Do we start accepting copying more widely in culture? Can projects like Factum point the way to a new, Anthropocene-ready preservation practice? As landscapes change and the known world erodes, should we commence a project of aggregating communal memories of the landscapes that once were? Digital storytelling is one way to develop this last point.[28]

Interdisciplinary Openings

> The longer I have worked on this issue, the more my focus has been drawn from my traditional field of expertise—the modeling of energy-economy systems—to the disciplines of political science, public policy, behavioral economics, sociology, psychology, and global diplomacy . . .
>
> I now believe that people with my expertise must learn from these other disciplines so that we might integrate our knowledge of the energy-economy system with their knowledge of how people make personal and collective decisions, including how they respond to challenges to their worldviews. —MARC JACCARD[29]

So far, we have discussed disciplinary research approaches to climate change. At the same time, we must bear in mind that interdisciplinary work is also proceeding and will surely continue. Indeed, many of the disciplinary examples already noted crossed departmental boundaries, such as the Turkish philosopher writing about religion or the artist doing philosophy and gender studies. The Nieman article called for crossing journalism's domains in reporting, echoing what seems to be occurring in higher education. We can find a great deal of interdisciplinary work, so much so that we might anticipate climate change to be the biggest driver of interdisciplinary academia in a century.

To begin with, even basic scientific research into climate change often involves participants and methods from multiple academic disciplines. Researchers from earth science, chemistry, atmospheric sciences, and biogeochemistry authored a major paper establishing the longevity of carbon dioxide in the Earth's atmosphere. Authors of one effort to model natural strategies to reduce global warming came from biodiversity, physics, geography, forestry, and ecosystems. The discovery that the Earth's poles have shifted slightly due to climate change, as glaciers have melted and redistributed their mass, came from a team including faculty in hydrology, geography, and environmental engineering.[30]

Researching the impact of climate change on human health can involve multiple fields. For example, authors of one paper concerning allergic and asthmatic changes came from schools and programs devoted to environmental studies, botany, biology, medicine, aquatic ecology, aerobiology, and clinical mycology, not to mention pulmonology. This multidisciplinary practice appears in other topics as well. Projecting the impact of different degrees of global warming can involve researchers from earth sciences, emergency management, and numerical modeling. Studying how irrigated lands might dry out can involve faculty from earth science, food systems, soil sciences, geology, biology, and multiple branches of engineering. A Princeton University study offering scenarios wherein America can advance to a post-carbon footing included authors from biology, forestry, evolutionary biology, soil and crop sciences, and environmental studies, not to mention mechanical, aerospace, and financial engineering.[31]

Researching human responses to climate change seems to be inherently, even radically, interdisciplinary, bringing together the natural and social sciences with the humanities. An attempt to forecast the potential impact of global warming on human mortality integrates economics, environmental science, demographics, and policy analysis. Thinking through the impacts of climate stresses on food supplies necessitates moving from environmental science to economics, food systems, sociology, and political science (the last two to account for social and political instability). Considering how what we eat shapes food systems, one study starts from the emissions that alter atmospheric chemistry, then moves to culinary culture.[32]

Social justice analysis of climate change is often interdisciplinary. Examples abound. Gender studies meets government and climate change in a paper exploring the role of parliamentarians' gender identity on climate policy. A study of different genders' carbon footprints is the work of an ecologist, a physicist, and an environmental scientist. An essay on personal reproductive choices integrates environmental studies, women's studies, demographics, psychology, sociology, and demographics. Similarly, an examination of direct air capture possibilities linked engineering and environmental studies to food systems and global developmental justice.

A consideration of how training large AI datasets impacts marginalized communities moves from computer science to geopolitics and language:

> When we perform risk/benefit analyses of language technology, we must keep in mind how the risks and benefits are distributed, because they do not accrue to the same people. On the one hand, it is well documented in the literature on environmental racism that the negative effects of climate change are reaching and impacting the world's most marginalized communities first. Is it fair or just to ask, for example, that the residents of the Maldives (likely to be underwater by 2100) or the 800,000 people in Sudan affected by drastic floods pay the environmental price of training and deploying ever larger English [AI language models], when similar large-scale models aren't being produced for Dhivehi or Sudanese Arabic?[33]

Such interdisciplinary work may yield new disciplines, as has often occurred in the past century. Already there are calls for emerging fields, such as collapsology and ecomedia. We already have decades of work in the environmental humanities, and seem likely to see more. A quick check of Wikipedia's entry on that field reveals that:

> there are dozens of environmental humanities centers, programs, and institutions around the world. Some of the more prominent ones are the Rachel Carson Center for Environment and Society (RCC) at LMU Munich, the Center for Culture, History, and Environment (CHE) at the University of Wisconsin–Madison, The Center for Energy and Environmental Research in the Human Sciences at Rice University, the Penn Program in Environmental Humanities at the University of Pennsylvania, the Environmental Humanities Laboratory at KTH Royal Institute of Technology, The Greenhouse at the University of Stavanger, and the international Humanities for the Environment observatories.

A more developed interdisciplinary field, science studies, which examines how scientific enterprises function and how scientific thinking proceeds, has been transdisciplinary for decades, as even a casual glance at the work of Donna Haraway or Bruno Latour demonstrates. Some interdisciplinarily intensive fields, like design, do not lose their plural makeup when grappling with climate change, as we can see by surveying projects at Harvard University's design school. Alternatively, we could see a call for an applied and interdisciplinary new field aimed at understanding and "stewarding" collective human behavior as a "crisis discipline," given the intense urgency of the climate crisis problem, even if our assembled knowledge is not sufficient. Joseph B. Bak-Coleman, Mark Alfano, and Wolfram Barfuss describe such a discipline in these terms: "Crisis disciplines are distinct from other areas of urgent, evidenced-based research in their need to consider the degradation of an entire complex system—without a complete description of the system's dynamics. We feel that the study of human collective behavior must become the crisis discipline response to changes in our social dynamics." The authors of the 2021 article "Stewardship of Global Collective Behavior" urge us to consider the development of such a new field, one that can operate very quickly and that "will require a transdisciplinary approach and unprecedented collaboration between scientists across a wide range of academic disciplines." Note that this vision is an active one, calling for interventions within human civilization.[34]

Changes to Research

We should expect some fields that devote the highest efforts to the climate crisis to grow, expanding in staffing and output as their research grows ever more vital. The historical record suggests disciplines may also branch off into new subspecialties as they grow, such as environmental assessment out of environmental studies.[35] The obverse seems likely to occur as well. Academic units that fail to grapple with climate change and that cannot make a convincing bid for their contributions could suffer a political hit as a result, to the extent that global warming plays a significant role in enrollment and reputation. That in turn makes it harder for them to defend their positions against budget cuts and reallocations. Depending on institutional and local conditions over time, being crucial to the climate crisis may become the leading way we think of academic relevance.

If we recall that "being crucial" is often a subjective matter having more to do with for whom something is crucial and the nature of the speaker, we start recognizing another potential way academic research can change over the next two generations. According to contemporary postcolonial, feminist, antiracist, and disability critiques, too much research expresses biases, which both warp results and lead to unjust benefits for the privileged. Too many articles and monographs reflect a Western, educated, industrialized, rich, and democratic (or "WEIRD") positionality, excluding multiple strata of the human race.

The antiracist uprisings of 2020 brought racism and white supremacy into focus for academia, pointing to racist bias in historical and contemporary scholarship. Recent work has called for scholars to heed injustice in the construction of public health knowledge, a call for "epistemic justice," which readily speaks to many other fields. These critiques then connect with the drive for climate justice in allocating resources worldwide with an eye toward equity and responding to historical inequalities. To the extent that these pressures find positive responses in the academy, we should expect scholarly research to transform: being more global and less local, more engaged with the fullness of human experience, better attuned to structural biases. Those pressures and positive responses will also elicit critiques, resistance, and even violence, based on recent history. Such a research transformation will involve struggle.[36]

Academic research is likely to change in another way. Travel to research sites, conferences, and consultations has been a feature of academic life largely made feasible at scale by plentiful air travel, which expanded massively over the past century. However, as aircraft emit greenhouse gases in flight, calls have arisen to reduce air travel. The Swedish language now includes the noun *flygskam*, or "flight shame" in English, describing the social drive to convince people to take other forms of transportation or simply not to travel.

This has already impacted higher education. In 2016, the Nearly Carbon-Neutral Conference model appeared, starting to explore how to conduct research without enhancing global warming. It has since published a helpful guide to "offer[ing] a nearly carbon-neutral (NCN) conference alternative, which is completely free of cost, that can reduce these carbon footprints by a factor of 100 or more." Crucially, the NCN community linked climate costs to social justice in their approach: "Given the horrific environmental costs and inherently exclusionary nature of traditional conferences, the time has come to radically rethink this cornerstone practice of our profession." Their approach now includes prerecorded video presentations and online discussion among participants. They also saw this as an improvement in some ways: "On average, the [2016] pilot conferences' Q&A sessions generated three times more discussion than takes place at a traditional Q&A. A few sessions generated more than ten or fifteen times more, making clear that, while different from a traditional conference, meaningful personal interaction was not only possible, but in certain respects superior."[37]

In 2019, Sophia Kier-Byfield wrote of her conference experience in terms of climate impact, arguing that much travel—especially that of senior researchers—could be cut back without professional losses. It would be the morally right thing to do: "If we are to be truly ethical in our research practices, we need to confront the high environmental price of the international conference circuit, which includes emissions and the wasteful use of finite resources."[38]

It is possible that some in-person events were losing attendance numbers as the twenty-first century progressed, with more attendees participating online. The Modern Language Association's annual convention statistics reveal a peak in 2002, a step down in 2003–2008, then a steady decline through 2019. Some of this may be due to researchers in the MLA area being progressively less likely to afford an expensive conference as tenure-track lines and hires in general declined, but the online option remained and improved gradually, following the technology and its adoption. Also in 2019, the Societies for Cultural Anthropology and for Visual Anthropology joined forces to plan for a reduced travel conference the following year. This led to a complex proposal. Rather than simply switching everything online, their hybridized design combined online presentations and virtual engagement with in-person travel to local events:

> Like its previous iteration (Displacements 2018), Distribute 2020 will be virtual and distributed: virtual in that it will be anchored by a dedicated conference website streaming prerecorded multimedia panels; and distributed in that presenters and viewers from across the globe will participate in the conference via in-person local "nodes." Distribute 2020 will offer three full days of streamed audio-visual panels and in-person local nodes where participants can gather with others to view the conference and join in related activities like workshops, art exhibitions, and dinner salons. Our goal is a low-cost, highly accessible, carbon-neutral conference that might pave the way for rethinking the mega-conference model.

In a similar vein, the Comparative and International Education Society planned a 2020 virtual conference "to reduce the environmental costs of attending a conference (i.e., carbon footprint associated with air travel), as well as alleviate other barriers that hinder participation of CIES members in the annual conference."[39]

During that next year, the COVID-19 pandemic struck and gave all of academia an involuntary sense of what entirely online operations could look like, including scholarly research. Synchronous video talks and discussions, prerecorded presentations, asynchronous exchanges across email and social media, conversations through multiple apps: 2020 saw us realize, sometimes awkwardly, a virtual research world. This has certain advantages, as Kier-Byfield and the NCN anticipated. As Dickinson College historian Karl D. Qualls put it, somewhat tongue in cheek:

> **Benefits of @aseeestudies virtual conference**
>
> 1. Proper and cheap organic breakfast from my garden (except org eggs from farmers mkt)
> 2. Sweatpants
> 3. Can attend while hanging upside down on my lumbar decompression table
> 4. Typed chat questions MUCH BRIEFER and on point than too-often pontification and self-aggrandizement
> 5. Can pop "out" for a cuppa and not miss anything
> 6. Sit in the sun on my deck rather than a depressing conference room in a hotel
> 7. No elevators[40]

At the same time, this jury-rigged, even improvised, scholarly system had many problems. Scholars spoke of their frustration at not being able to interact with peers in person, finding technological mediation to be often flawed and restrictive. Many faculty saw their research agendas paused as they could not physically access key tools and materials. Women researchers in particular saw their work suspended as they took up the lion's share of home and care duties. Personally, I encountered a great deal of pushback when conducting research on this topic. Simply raising the idea in various venues—not advocating for it but rather offering reduced faculty air travel as a topic for discussion—elicited sadness, protests, rancor, and even insults. Some academics argued that a *flygskam*/NCN event would be anti-intellectual or anti-academic as it would devalue crucial social connections among colleagues. In another view, academic travel had miniscule environmental costs, and academic energies would be better exercised on options with larger impact. Others saw reducing faculty and staff travel as especially damaging to early career researchers, who might not be able to network and establish themselves sufficiently. Another charge took aim at the hybrid Societies for Cultural Anthropology and for Visual Anthropology conference style, seeing that it penalized people working in institutions too isolated to access local, in-person events. One leading librarian observed that universities tracking faculty and staff professional development gave more weight to in-person than online presentations.[41]

How can we decarbonize academic travel, learn from our collective COVID experience, benefit from the innovative work of the NCN and others, and also solve all of these challenges? This is one design problem higher education will have to start to address quickly with the pandemic's lessons fresh in mind and before contributing even more GHG to the world. We will return to this topic in chapter 7.

One reason for academic travel brings up another research challenge. Some faculty journey to locations in order to do research: archives whose contents are not yet digitized, fieldwork, and so on. Those sites may prove challenging to access as the climate crisis persists. An archive might lock down against weather or lock up for purposes of relocation. Areas for fieldwork can be compromised by different aspects of changing climate. Getting ice core samples becomes problematic when the cryosphere moves and melts. Rivers, lakes, and oceans change in many ways, as previously discussed, from heating up to becoming more acidic. An American research team identified striking geological features on an Atlantic site, only to discover their impending doom via warming waters and ocean acidification.[42] Conversely, parts of the Atlantic Ocean could become much colder if the AMOC currents slow significantly or, worse, stall. Such changes are subject to research, of course, but also represent a challenge to long-running projects. Historical sites may be restricted, restructured, or relocated. Rising climate change could incentivize researchers to get their fieldwork done as soon as possible, or else face delays, dead ends, or alterations to their research materials.

Additionally, some researchers will rethink the material apparatus of their work in order to reduce its greenhouse gas footprint. We can see an early sign of that in the story of one Dartmouth College climatologist:

> Osterberg wanted to cut his environmental impact when drilling ice cores. He was delighted when, in 2012, he got permission to drill inside Alaska's Denali National Park—but he wanted to avoid using a petrol-powered generator. So he asked engineers who design drills for US coring operations to develop a low-impact rig. They delivered. In 2013, Osterberg's team successfully drilled two 200-plus-metre cores by drawing power from only a few solar panels, a wind turbine and a bank of batteries.

Such changes can also save money, which may drive further research. Administrators may urge this kind of research greening on faculty for combined pecuniary and environmental reasons.

Research can be critical to faculty tenure, and that institution may change as the climate crisis persists. If interdisciplinarity becomes crucial to research into the full breadth of global warming, then universities, schools, programs, and departments will have to alter their tenure and promotion structures in order to support it. As one report put it, "To effectively address the [climate] adaptation challenge—which will require integrating many disciplines such as climate science, economics, public policy, psychology, engineering, and more—research institutions will need to identify and reduce the structural barriers to cross-disciplinary research, revising policies to incentivize and reward collaboration across disciplines. In addition, they will need to examine the challenges faced by interdisciplinary teams working across multiple universities." Professor and administrator H. Tuba Özkan-Haller adds that faculty will need support in doing public communications work about climate research, along with outreach to marginalized communities. The tenure review process should recognize and support this.[43]

To sum up: academic research plays a crucial role in empowering human understanding of the unfolding climate crisis. While much work occurs within disciplinary housings, interdisciplinary and transdisciplinary research is rising. Looking decades ahead, we can imagine an academy increasingly devoted to studying the Anthropocene through familiar academic departments and new ones. Depending on an institution's nature and strategy, we may forecast new forms for such work: centers, institutes, academic programs, colleges within universities, entire standalone colleges and universities devoted to researching the topic. Connections across institutional barriers may be feasible, from public-private partnerships to multi-institutional networks, citizen science, and crowdsourcing projects, such as an effort by the public to transcribe ship logs in order to track climate patterns as the world changes.[44] Yet politics, including politics driven by climate change, can stymie such connections.

Supporting some of this research will become challenging at times. Climate stresses can damage or block access to research materials. Political pressure can press against academic freedom. Financial stresses can vitiate a research agenda. Further, not all faculty will be organically ready to advance research into the Anthropocene, which means professional development will be required. Campuses might convene interdisciplinary groups to boost faculty climate change research work. This could involve bringing in outside speakers, conducting collaborative research, sharing work in progress, or introducing or training on key tools.

All of this chapter's discussion has concerned research production. How the creation of new knowledge about the unfolding climate crisis plays out in teaching—through curricula and pedagogy—is the subject of the next.

CHAPTER 3

Teaching to the End of the World

We need to speed up our cycles of learning. That is, we must understand that not only are we in a discontinuity, but we're unlikely to emerge from that discontinuity into a stable state. There is no new normal coming. We are going to experience a sort of standing wave of discontinuity, with no insight ever becoming certain, no expertise ever being quite definitive, no strategy ever being completely reliable. We're going to have to keep relearning, over and over again. That is simply one of the costs of having destabilized the planet. —ALEX STEFFEN

Seems there's a functioning university now in Vancouver. They got back on their feet faster than anyone expected. And they're looking for bright students who want to learn engineering and philosophy and applied sociology and all the other things that can rebuild this broken world. —HOLLY SCHOFIELD, "HALPS' PROMISE"

We must rethink our educational curricula if we are to recapture and sustain the influence so necessary to shaping a better world. Just as race, class, gender, and sexuality studies came to pervade humanities research and pedagogy during the past three decades, it is imperative that ecological studies also be integrated in similarly prominent ways. All connect the individual to the collective in addressing historical and present problems and all seek solutions for wider inclusion and justice. Indeed the threat of climate change is an existential crisis, a crossroads of conviction about how we define our personal identities, not in isolation, but in recognition of the intricate web of relationships that sustain the earth we inhabit. —ROBERT D. NEWMAN

Now that we've explored how the climate crisis could impact academic research, we can turn our attention to what many consider to be *the* essential function of colleges and universities: teaching. In this chapter, we begin by considering the futures of curricula and pedagogies. Next, we address other dimensions of academic education, including non-classroom life and potential expansions to the student body.

Curricula

To begin with, what does a climate crisis curriculum contain? What topics does it teach?

Walter Leal offers a comprehensive survey:

> The development of a broader knowledge among university students about the basic principles of climate change and the degree of contribution provided by the natural sciences, social sciences, economics, architecture, etc., towards both the understanding of climate change and the complexity of problem-solving approaches related to it.
>
> The education of university students in the socio-economic issues (e.g., poverty, social justice, security) associated with climate change and to which governments need to find a solution in order to ensure the survival of people and habitats.
>
> The motivation of students to take action both during their time as students and, later on, as professionals.

We can note several key curricular themes in this passage. First, the importance of individual disciplines in teaching from their particular frameworks. Second, the role of interdisciplinary connections, especially with an emphasis on the social sciences. Third, the power of activism, both in the public sphere and on campus.

This much is recognizable in the present. We may also expect changes to research to drive curricular developments in the future. This is partly due to historical connections between research and curricula, as institutions often select topics for study on both levels based on a general sense of their value—that is, we value mathematics both for students to attain a degree of numeracy and for supporting new ideas about number theory. Further, some faculty introduce their research into their teaching, depending on their institutional culture, departmental position, and personal inclination. Additionally, scholarly work on education and learning can inform how colleges and universities determine what they will offer in classes.

We can also view higher education's climate-related curricula through a supply and demand lens. Student interest in the topic can drive expanded amounts of classroom content depending on institutional flexibility and responsiveness.[1] Also on the demand side, faculty interest can add to the amount of curricula centered on the climate crisis, again depending on faculty roles and institutional culture. Faculty also work on the supply side of the question, of course, providing more content for study. College and university systems enhance this through program design, graduation requirements, and cocurricular programming. Off-campus organizations can also supply climate curricula in the form of content: learning materials produced by scholars and other interested parties on their own or through publishers.

Therefore, we should expect increased research on the climate crisis to yield rising amounts of teaching about the subject. When campuses choose to see the Anthropocene as a subject worthy of publication, they are likely to add sections, classes, or programs on the topic. As individual faculty bring their research into their teaching, student demand may also rise. Curricula may also respond to external pressures. Local or national governments may request or mandate more coursework on climate change. They can also push back, discouraging or banning such teaching. Funders, both private and public, can also incentivize more climate teaching. More broadly, climate developments in the world could encourage or discourage academic units from expanding relevant curricula, such as major natural disasters that academics interpret as caused wholly or partially by climate change. Fossil fuel companies may seek to influence learning content as they once did in primary and secondary schools.[2]

In terms of specific academic programs, climate change content may follow the research coverage we discussed in chapter 2. Many of the natural sciences can—and sometimes already do—teach entire classes or programs on climate issues. Environmental studies is the obvious candidate here, as well as oceanography, hydrology, meteorology, and civil engineering. For example, Caltech's Climate Dynamics program offers a series of classes on atmospheric measurement and systems.[3] The social sciences can also teach the climate crisis, as economics, psychology, anthropology, and others can each offer its perspectives on how humanity grapples with the problem. Similarly, the humanities can teach climate through its lenses, looking at global warming as a matter for historical, philosophical, literary, or artistic consideration. Each can offer classes explicitly concerned with, or themed to, climate change, such as weather systems in the Anthropocene, climate change ethics, or the psychology of global warming. Each can also use the topic as one within other classes. A literature seminar on the British Romantics can have a unit or theme on how those writers depicted climate crises, reading Byron's "Darkness" (1816), Percy Shelley's *Revolt of Islam* (1818), and Mary Shelley's *The Last Man* (1826). Students in a civil engineering capstone class may have the option of designing seawalls.

Beyond individual classes, we can expect to see changes to departments and programs. An academic discipline might collectively decide to devote more teaching to climate change as a matter of public good if not civilizational survival. Such a decision could rely on projecting changes to the professions a given discipline tends to support. For example, Emily Atkin called for journalism to consider climate reporting central to all journalists: "We understand that [climate change] infiltrates every single area of our life . . . There is no excuse for a reporter today who doesn't understand the basic science of Covid-19. Why is it not the same for climate change? Everyone should be a climate reporter. And if you are not a climate reporter right now, you will be."[4]

Elsewhere, statistics and mathematics may introduce more climate curricula as those fields prepare students for careers in insurance, an industry that grapples with global transformation and risk in a major way. Teacher education programs, which prepare students to teach in primary and secondary schools, are also likely to change as those educational institutions redesign their own curricula for the unfolding crisis. Already, debates over whether and how textbooks describe global warming are under way. We should surely expect related discussions within postsecondary classrooms and departments.[5]

Individual campuses may also offer new programs in various forms at the undergraduate level. The United States has developed many examples of such across a range of institutions, including full degrees in various forms. The University of Colorado Boulder offers a major in environmental design. Pacific Union College launched a major in conservation technology, which includes course work in geographic information systems (GIS), biology, emergency management, and data science. The University of New Hampshire now makes a double major available, one that allows students to pair a major in sustainability with another major. The University of California, Davis offers a Sustainable Environmental Design major. Note its stated mission:

> Urban planners and regional developers shape the physical landscapes of our communities to create environments that frame our daily lives. With an added focus on sustainability, these community builders help ensure that the places in which we live and work are environmentally safe and resource-efficient. The sustainable environmental design major prepares students to thoughtfully plan communities, design livable cities, develop smooth-running transportation systems and create healthy, equitable urban spaces.[6]

It doesn't address climate change per se, but it clearly works to prepare students to help design cities that can best survive the unfolding crisis.

Some campuses offer minors in and around the climate crisis. The University of Maryland launched a Sustainability Studies minor. The University of Washington offers a Climate Science minor. We can think of students taking these to complement or expand on their majors. A student majoring in business, say, can inflect that course of study with the sustainability minor, combining learning how to drive an enterprise forward with how to do so in a green fashion.[7]

Graduate programs are also growing around the world. In Europe, Hamburg University of Applied Sciences supports PhD work in its

Sustainable Development and Climate Change Management Research and Transfer Centre. Master's degrees at a range of institutions offer different approaches to the topic. For example, Erasmus University Rotterdam has an Urban Environment, Sustainability and Climate Change program, while Victoria University of Wellington's Master of Climate Change Science and Policy doesn't have quite the urban focus of the former and might more closely resemble the University of Groningen's MS in Cultural Geography–Climate Adaptation Governance. Lund University's MS focuses on risk management, while Swansea University's MS emphasizes cartographic visualization. Columbia University offers a master of arts degree in Climate and Society, based in its Climate School. Some of these graduate programs are based in institutes, as in the University of Maine, which hosts an MS in Quaternary and Climate Studies in its Climate Change Institute.[8]

Institutions can add some measure of climate study or awareness to general education requirements. Think of a college where all students must take at least one class in the climate crisis—it could be in any department, so long as the campus's chief academic officer approves. Perhaps others will think of some form of *climate literacy* as a graduation requirement. Students could fulfill this in many ways: individual classes, a string of programs, or by passing an exam. The University of Connecticut already offers "e-courses" for its environmental literacy requirement.[9]

We could also see campuses requiring green physical work for a degree. "Work colleges" in the United States already include some hands-on labor in their curricula. We could imagine requiring students to spend a certain number of hours working on seawalls, insulation installation, forest management, or digging up paved roads for replacement. One example of this comes from the Philippines, where a 2016 law requires each college graduate (plus high school graduates) to plant one tree apiece. Notably, the sponsoring legislator framed this measure in terms of intergenerational equity. The nation's government plays an administrative role: "The country's Education Department is responsible for implementing the new rule, while others, such as the Environment and Agriculture Departments, will establish nurseries, supply seedlings, identify suitable planting sites, and monitor the trees' progress."[10] We can easily envision colleges and universities around the world implementing such requirements on their own. Nations and provinces, states and cities could mandate the same, especially for public institutions.

Yet as we saw with academic research, academic teaching into climate change is not likely to be solely the province of individual disciplines. Just as investigating Anthropocenic matters can cross departmental boundaries, so helping students learn about global warming may often be an inter- or transdisciplinary affair. A student curious about how changing climate impacts their home city can easily follow paths laid down by environmental science (what do rising temperature do?), atmospheric science (will we be wetter or drier?), hydrology (how high might the waters rise?), sociology (how will people react?), and health care (will diseases run rampant?). Team-taught classes featuring faculty from several disciplines are one way to support such an approach to learning. Interdisciplinary centers may also prove to be useful resources, anchors for new classes, as with the Center for Climate Change at Ithaca College. Inter-campus digital resources for how to teach climate change across departments are already being used, like the Teaching Climate Change in the Humanities website.

In addition to digital resources, the physical campus itself can serve as a learning resource. As we saw in chapter 1, many colleges and universities will rethink and redesign their built environment and natural setting. That process can provide numerous opportunities for students to study, work with, and participate in faculty research. Creating new buildings, installing wind turbines, setting up botanical preserves, and planning on environmental changes are all potential materials for learning.

Liberal arts programs and institutions may be particularly well suited to support such interdisciplinary work, given the discipline-crossing aspect of their tradition. Perhaps the climate crisis will be a new driver for student and institutional interest in the liberal arts. Alternatively, some may consider dealing with climate change to *be* "the new liberal arts."

What strengthens this assertion is awareness of another part of that tradition, as *artes liberales*, the necessary preparation for a free person to engage and prosper in the world. Since that world is being reforged in the Anthropocene's fires, it calls for a new liberal education indeed.

Embodying this liberal climate arts approach might be a multidisciplinary and multi-instructor class about climate change. Imagine one in which a different faculty member teaches a class each week from their professional perspective. A physicist begins the semester with a description of how the earth's systems interact with solar radiation. A geographer leads a class that breaks down the global details of global warming. A meteorologist breaks down the global weather systems. Then a political scientist leads a week on the ways humanity has and can organize in the climate crisis. A sociologist follows from their perspective. A psychologist works through the mental implications, and so on. Perhaps the penultimate class is the domain of instructors in writing or communication, who help students compose final projects, while presentations of that work go live in the institution's library or online. Such a class would immerse students in the breadth of the problem while introducing them to faculty members and academic disciplines. Faculty members can attend sessions for professional development and collaborative purposes.

All of these efforts can occur at a single institution. One college or university could host a climate studies center, consider the topic to be a new way of implementing liberal education, team teach certain classes, and also teach the crisis within individual departments. We can also conduct classes hosted by, and including students enrolled in, multiple campuses simultaneously. This can let faculty who are specialists in one narrow part of their discipline partner with those in other domains, creating a curricular synthesis. Examples of this practice on topics other than climate change occurred dating back to the 1990s, from the Sunoikisis "virtual classics department" to senior humanities seminars shared by clusters of independent colleges to an interdisciplinary course on the American war in Vietnam that I cocreated and cotaught. More recently, dozens of colleges in the Council of Independent Colleges (CIC) shared upper-level humanities seminars with one another.

Colleges and universities may expand their sense not only of *what* to teach but *when* as they commit more seriously to lifelong learning. This is the idea that students may turn to academic institutions throughout their lives as their careers and interests develop. The sixty-year curriculum concept offers a good example of this, as does the Open Loop University concept from Stanford's design school:

> Upon enrollment, students received six years of access to residential learning opportunities to distribute across their lives as they saw fit.
>
> Many students chose to concentrate their on-campus stint for a few years on the earlier side, as the social process of maturation within a peer group remained important. Others, freed from social stigma, attached to taking gap years or years "off" during their educations.[11]

Instead of just focusing on a single academic experience for each student, often early in their careers, a postsecondary institution offers learning in various ways throughout a student's working or even biological lifespan. The unfolding of the climate crisis over decades could drive many such lifelong learning reforms.[12]

There are many ways this could play out. For example, a student may spend five years as an undergraduate studying topics other than the Anthropocene, then return twenty years later to take a certificate on mitigating climate change. A graduate student in public health could focus on how global warming drives new forms of and sites for disease, then return in five years for an associate's degree in sociology in order to better understand human reactions. Retirees may take classes to understand the transformed world, while high school students sign up for postsecondary classes that go beyond what their current school offers.

Climate change can also impact lifelong learning operations by encouraging or actively driving adults out of GHG-intensive jobs. An alma mater, or simply a university with an especially competitive adult education initiative, may prove essential to such career switchers. Some colleges and universities may win large numbers of learners by offering to guide them into exciting post-carbon jobs. Put another way, we may see the emergence of new curricula for a post-carbon world.[13]

Pedagogy

Curriculum is the domain of *what* is taught. Now we can explore climate-driven changes to pedagogy, or *how* we teach. As climate change transforms our curricula, research, and campuses, so it may also revise our teaching and learning practices.

The COVID-19 pandemic experience may give us some glimpses of future teaching, concerning the mix of face-to-face and online education. As we saw during lockdowns and quarantines, we may anticipate more remote learning if environmental stresses make in-person learning difficult or impossible. Dangerous heat waves, floods, fires, and, as we've seen, disease outbreaks can switch education into the wholly online track quickly. Problems with transportation networks may have a similar effect. As we saw in 2020, this means thinking ahead about different aspects of instructional continuity. Not only do we have to solve once more for problems of connection and access, including digital divide and equity issues, but we may have to rethink assessment, student workload expectations, and mental health challenges for each student, not to mention helping them with their physical safety as needed. Between the two polar options of in-person and entirely online learning, we can also expect various forms of what Brian Beatty memorably named HyFlex teaching. With this style of instruction, individual students, faculty, and support staff can choose if they will participate in person or digitally for a given class meeting. COVID showed that conducting such blended online and face-to-face experiences is doable, if challenging in terms of hardware, infrastructure, instructor preparation, and student practice. If climate pressures make it difficult for members of a class to convene in person, or if academia decides to integrate online and face-to-face learning within individual classes, or both, HyFlex may become the twenty-first century's baseline teaching mode.[14]

Additionally, there is a world of opportunity for teaching in locations outside the physical classroom. More hands-on work in nature, from forests to rivers and caves, can ground students in some climate subjects better than more conceptual or theoretical work within classrooms.[15]

New Orleans professors Ned Randolph and Christopher Schaberg call for drawing student attention to the climate evidence and stories right outside their campuses:

> Teaching environmental awareness on the bank of the Mississippi River, we can reflect on its eons of meandering channels and more recent attempts to control its course; our close proximity to the river offers direct lessons in hubris and the ecosystem's long duration. Nonnative species are not abstract subjects but nearby entities: from the brown anole lizards skittering everywhere to the *Triadica sebifera* tallow trees erupting from sidewalks, we can see the effects and tensions of global migrations right around us.

Some of that evidence and some of those stories are historical and cultural:

> We can translate this attention to the cultural landscape around us, too—the dense weave of narratives and traditions, threads of violence and resistance. They can be found in the muddy swamps around New Orleans that offered tactical refuge to Native Americans and escaped slaves in "maroon" communities throughout the 18th and 19th centuries. Indeed, we see this historic relationship emblematized by African Americans who hand-stitch elaborate Mardi Gras Indian outfits to pay homage to this deep relationship. Our students, likewise, need only look up or downriver to witness the cultural and economic connections between modern petrochemical industries in Louisiana and the plantation system from which they were born. Today, extractive, dehumanizing practices of these industrial plants that stand on former sugarcane plantations pollute the water and air of the historic black communities that surround them.[16]

More specific assignments may include those based on memory and environment, such as asking students to memorialize and reflect on certain landscapes before they are replaced. There is an analog to HyFlex in that students may have access to digital networks in some (if not all) locations, giving them the ability to upload captured media, query the open web or databases, and work on projects. To make more place-based learning happen, we may consider changing assignments or setting up more physical sites readily available for teaching, such as arboretums.

Instructors may rethink the role of group work. Currently, classes rely on this pedagogy for a variety of reasons, including helping students develop teamwork skills for the work environment. Climate change adds other reasons. If some faculty think that forms of individualism helped power human society into a greenhouse gas disaster, they might want to de-emphasize individualism in the classroom and shift student learning into more of a collective process. Other instructors might want students to learn from each other's climate experience and therefore structure small group discussions along those lines.

There are many ways digital technologies can support climate change learning. We can start with the panoply of devices, software, and practices currently available, if not always used: learning management systems / virtual learning environments, synchronous videoconferencing, lecture capture, wikis, presentation applications, and many discipline-specific tools, from mapping software to statistics packages. Mixtures of these can enhance in-person classrooms and expand learning between class sessions. Augmented reality (AR), the addition of digital content to a physical location, can be a powerful visualization tool. Meteorology has already used AR to superimpose rising waters on landscapes. Mount Resilience, an Australian Commonwealth Scientific and Industrial Research Organisation (CSIRO)–Australian Broadcasting Corporation project, superimposed elements of communities in climate crisis on participants' smartphones and other devices. This kind of experience can make the potential experience more vivid to people, bringing it into their personal spaces in an interactive and safe way. We can imagine mobile apps that represent a storm striking the local campus, or new buildings on a present-day landscape.[17]

Digital storytelling projects may give students spaces to reflect on the climate transformation, including centering their own experiences. Digital storytelling can also help students develop their personal voices, a personal link that may be both humane and vital in a time of global-scale challenges. Students can capture media (images, sounds, videos) about climate change, synthesize those documents into a structure through their classwork, then give it narrative form by turning it into a personal story. They can also use their personal experience as story material, integrated with their studies. For example, Kelly Hydrick connected her childhood memories of bees in nature to her children's experience of the world, framed by her historical studies. Faculty and staff can also create digital stories based on their personal lives or professional work. One example of this is the climate migrants story map produced by faculty working in the international Trans-disciplinary Research Oriented Pedagogy for Improving Climate Studies and Understanding (TROP ICSU) project.[18]

Another pedagogical practice that can be of use in teaching the climate crisis is gaming and simulations. The general capacity of gaming for learning has been established for two decades and does not need recapitulation here. The particular use of gaming for teaching climate change does require some explanation. To begin with, games can portray complex systems in operation, such as ecosystems or the hydrologic cycle or cities, giving students a view of them along with ways to intervene in their process. Such artificial constructs can be reset and replayed or rerun to try and test different strategies—this is even more effective with digital games. In addition, role-playing games allow students to take up different human decision-making and experiential positions within a system. La Universidad Autónoma de Madrid has run climate crisis role-playing games for years, starting in 2001, with games representing "the Kyoto Protocol negotiations . . . the Rio+10 Summit's negotiations . . . young people's strategy for participation in environmental themes of the UN . . . [and] 'Who Owes Who?' Campaign," a "tribunal to judge the legitimacy of the external debt."[19]

For example, Climate Interactive (CI) has produced two simulation tools. C-ROADS simulates the impact of different national and regional decisions on the global climate through the year 2100. Users can adjust various values (the year when GHG emissions peak, annual rate of emissions reductions, deforestation and afforestation rates) for major nations and regions (China, the European Union, India, the United States, two collections of other countries) and quickly see the impact of those choices on global temperature increases and overall net GHG emissions.

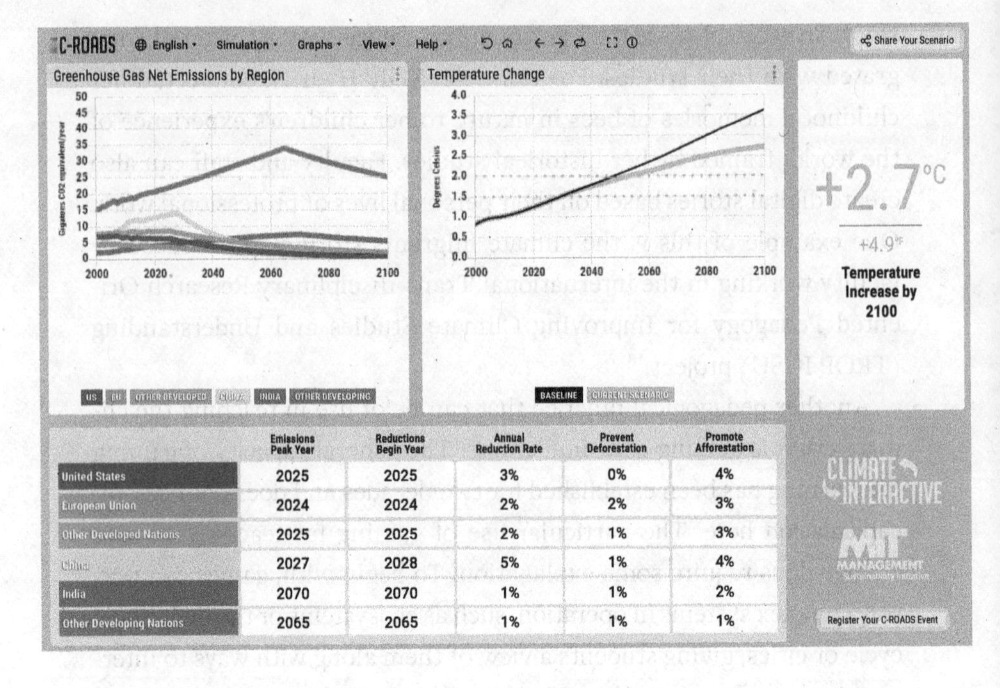

	Emissions Peak Year	Reductions Begin Year	Annual Reduction Rate	Prevent Deforestation	Promote Afforestation
United States	2025	2025	3%	0%	4%
European Union	2024	2024	2%	2%	3%
Other Developed Nations	2025	2025	2%	1%	3%
China	2027	2028	5%	1%	4%
India	2070	2070	1%	1%	2%
Other Developing Nations	2065	2065	1%	1%	1%

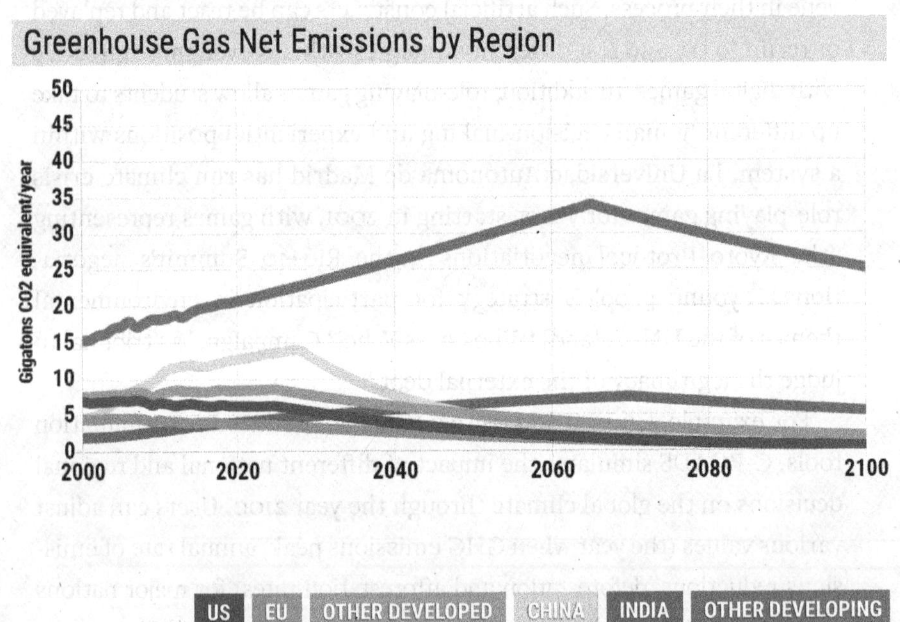

(*continued*)

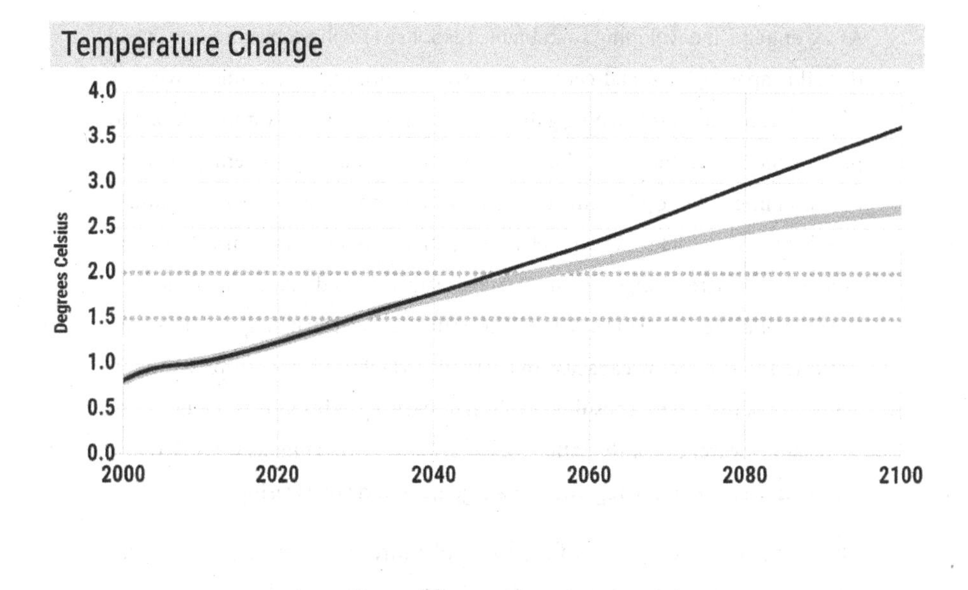

An example of C-ROADS in play, with emissions declining over time yet not enough to keep warming below 2 degrees Celsius. Climate Interactive

A student can rerun this basic simulation and enter different data to check results.

CI also created an in-person role-playing game, the World Climate Simulation, for "groups of 8–50 people (although it has been adapted for as many as 500) . . . A facilitator plays the role of a UN leader, while each participant plays the role of a delegate representing a specific nation, negotiating bloc, or, in some cases, an interest group. Everyone then works together to reach a global agreement that successfully keeps climate change well below 2°C."[20]

Another example is EcoChains, an academically created game that appears in both web and tabletop versions. It teaches players about the relationships between life forms in the Arctic region and how climate change tears at that complex web. One scholarly assessment of EcoChains found it taught systems thinking with results comparable to reading an article on the topic:

> As we engage in developing sustainable responses to climate change, we need to build capacity for social-ecological systems thinking . . . [G]ame players demonstrated similar learning gains regarding the climate system, actions to protect the Arctic, human actions causing environmental problems, and the base and mid-level food chain. This is important because games reach different people than do articles. As we seek to broaden discourse on social-ecological aspects of climate change beyond those already engaged (Turrin et al. 2020, Pfirman et al. 2021), even the ca. 50% overall systems learning gains from the game add to enhancing capacity. In particular, the fact that game play and active participation in it resulted in similar learning gains as reading in increasing awareness of the influence of actions is important as we seek to inform decision making regarding management and stewardship.[21]

Other games offer views of undoing climate catastrophe. *Wayfinder*, a National Film Board of Canada game, positions the player as an agent of ecological restoration, moving through an abstract landscape to gradually recover a sense of the world before climate change. Forthcoming game *Terra Nil* reverses a classic game genre, the city-builder, where players assemble a growing urban area over time, bit by bit, system by system. Instead, *Terra Nil* lets the user disassemble an emissions-heavy city, greening it by adding new plants and mammals, direct air capture installations, and post-carbon human structures.[22]

It seems likely that we'll also see climate change appearing in games not explicitly focused on addressing global warming. Tabletop and computer games have for years included current events in their content, much like other media do. For example, the designer of *In Other Waters* (2020) stated that the climate crisis was fundamental to his conception of the game, and reviewers have detected that.[23] Discussing games in class, students may start thinking about climate change in their broader gaming experience.

Many of these pedagogies challenge other pedagogies. Lecture-focused teaching presents a very different method, and instructors habituated to it would benefit from institutional and external professional support in order to branch out. Test-focused classes will need to adjust to account for the preceding methods, either by reducing assessments or revising them to capture learning that occurs on more constructivist levels. Fortunately, neither professional transformation is novel, and we have decades of scholarship and practice to bring to bear on it.

We may also help students cope with certain digital information problems, including information overload and disinformation as they apply to climate issues. The digital, information, and media literacy fields are very important on this score in that they develop students' abilities to discover, assess the quality of information, and make productive use of it. Media, information, and digital literacy units can be useful in classes as they fit curricular needs. For example, an information literacy module in a virology seminar may help students avoid poor information about climate-caused diseases.[24] We can also structure assignments to take students through a challenging information landscape, such as having students produce map-based story visualizations. Digital storytelling can also play a role in helping students grapple with information floods through the process of selecting and composing information within a narrative framework.

On a different register, teaching about a civilizational crisis can cause problems when this surfaces student anxieties within the pedagogical experience. Climate topics can challenge students emotionally or trigger traumas, including those that a student identifies with climate change. Simply reacting to the magnitude and scope of the climate crisis can elicit painful student reactions, which the instructor has to address. We have already seen examples of this in classes teaching the climate crisis. To pick one instance from a philosophy course:

> In 2018 [one professor] informed his ethics class that the amount of carbon in the atmosphere was over 400 parts per million, and that there was no hope of getting it down to 350. In the back of the class, a student started crying. "If I didn't have hope, how could I live?" she asked.
>
> [The instructor] had an urge to say *Exactly*. But he felt bad about making her cry.

As one recent study put it, "Climate change has significant implications for the health and futures of children and young people, yet they have little power to limit its harm, making them vulnerable to increased climate anxiety."[25]

Additionally, to the extent that climate has impacted their lives, students—as well as instructors and support staff—may also bring to class trauma unrelated to the specific classroom topic. That is, the Anthropocene will damage some students before they appear in class. Faculty members and support staff may also experience various forms of climate-related trauma and anxiety, of course. This can make us depressed, gloomy, or just resigned to a grim future. As Schaberg and Randolph describe their own experience, "In some ways, residing and teaching here [New Orleans] right now can feel fatalistic. Like we have to teach through the doom."[26] The COVID-19 pandemic demonstrated one pedagogical response: instructors being open, even vulnerable about their mental status with students, making themselves more available for emotional connections with them. Perhaps that kind of personal openness will work for some instructors during the Anthropocene.

At the same time, instructors need to take additional care with their students' mental health. To do so effectively could mean partnering with institutional counseling services, taking professional development programming to build up understanding, using content warnings or sensitivity readers in syllabus creation, partnering with off-campus (including online) resources and organizations, and more.

One way to support students experiencing psychological problems during climate crisis curricula is to give them opportunities to express positive visions of upcoming decades connected with a sense of their own agency. This is the opposite of setting students up to feel powerless in the grip of vast and deleterious forces. Some early research has shown promising results along these lines. An Iranian group working with younger student projects found that students who saw themselves participating in the future in an active way tended to create positive, even utopian images of what might come next. Perhaps a similar classroom approach could serve well in the undergraduate experience.[27]

Another way to support suffering students is to develop and offer classes designed precisely for that goal. One example comes from the University of Connecticut, where professor Robert Thorson teaches The Human Epoch: Living in the Anthropocene (GSCI 1000E) to first-year undergraduates. The class starts by defining and explaining the Anthropocene, situating it in the long, long context of geological history as well as extending it into the future to help students imagine what might come next. Professor Thorson observed that "students feel less neurotic after class because they have more knowledge." That understanding undermined any fear of the unknown. In a related example, a European university empowered students to create a class about a recently released IPCC report. It succeeded, working with faculty and producing multimedia documents for the class.[28]

Academic faculty may also teach openly optimistic content. Positive visions of the future from science fiction, seen through text, video, music, art, and games, may inspire students to think of climate futures they can help create that are not horrific and that summon up their creative and analytic capabilities. I have in mind the emerging solarpunk subgenre of science fiction, which imagines worlds wracked by climate change yet wherein people have carved out spaces that are more just, humane, progressive, and open to possibility. As Isaijah Johnson puts it, solarpunk may work in classes both through consuming and producing it:

> Solarpunk fiction can be considered pedagogical: through depiction and sparking the imagination, the knowledge, skills, and creativity necessary for producing an ecological future make their way into the hearts and minds of readers. This is a virtue which can be brought out either through reading solarpunk fiction or writing it—indeed, writing solarpunk demands that students consider what aspects of the socio-economic structure contribute to environmental harm and imagine how they can be overcome or improved.[29]

As the twenty-first century proceeds, we may see solarpunk flower into a major artistic current. It may be succeeded by others. The key point is that teaching optimistic topics within the category of climate change may empower students.

Teaching outside the Classroom

While colleges and universities emphasize teaching and learning within classes—and rightly so—students can also learn in and around campus spaces beyond the physical or virtual classroom. Historically, institutions have hosted visiting speakers, performances, movies, and other visitations, each of which can engage, enlighten, baffle, or infuriate students, who can attend these on their own. Campuses can also be home to museums, libraries, observatories, arboretums, businesses, nonprofits, farms, and other sites for additional learning. The act of being present with other human beings is, of course, a potential learning experience, especially for younger learners who may be living partially or entirely on their own for the first time. We can think of these as cocurricular opportunities or, in the United States, elements of "residential life." The point is that, taken together, colleges and universities can offer additional ways of learning about the climate crisis outside of, or adjacent to, their classrooms. We can easily expect institutions to offer speakers, presentations, movies, exhibits, and so forth that offer information and perspectives about global warming, with the range of responses (illumination, bafflement, etc.) noted previously. Campus leaders may seek to select such extramural content strategically. Students may also play a role in programming such.

Beyond museums and speakers, campuses also provide services linking students to jobs, from career fairs to recruiters and internships. If an institution wants to expand its climate engagement, it can do so along these lines. Think of a green jobs fair or a presentation called "Fight Climate Change while Making a Living." Students could undertake internships with relevant enterprises, such as climate think tanks or seawall construction projects. Some fire departments and firefighting associations are interested in technological innovation to improve their services, which should surely appeal to some students. Such partnerships can combine hands-on work with classroom theories, adding service time and job preparation to boot.[30]

For students who live on a campus, there are additional options for climate education. A school may offer climate-themed housing, such as food gardens attached to, on top of, or within dormitories. Students interested in reducing their overall GHG footprint may find housing dedicated to simpler living appealing. Collaborative housing projects involving people sharing food preparation, transportation, and cleaning may similarly appeal to some. Further, one public educator suggests another angle: "Maybe the model of student-run student services will become more common, for financial sustainability and resilience but also because maybe we will recognize that gardening and the liberal arts are not anathema to each other."[31]

Some students may learn outside the classroom from political activity anchored on their campuses, following a long historical tradition. They learn both intellectual content and the practical knowledge of politics through participating in fundraising, protests, campaigns, and more. Climate activism already energizes some students, who in turn give energies to groups like Extinction Rebellion. We should expect more along these lines, including targeting campus activities. We've already discussed students contributing to college or university divestment campaigns. We could see students demanding other changes on campus, such as altering food service, syllabi, career services, and institutional operations. They could demand an end to academic positions funded by oil companies, or protest research and classes that do not (in their estimation) truly take the climate crisis seriously enough. Such activism could extend to a broader sense of the campus community if students consider what their institution's graduates have done. For example, how will students think about alumni who work at carbon industry firms? Students can target certain aspects of campus operations in order to win attention or galvanize support, as when student groups from two American universities delayed an athletics event.[32] We can imagine student movements targeting rectors, trustees, government officials, and donors. Not being on a university's formal curriculum, and perhaps being opposed by even a majority of faculty, doesn't mean such events are not learning experiences.

Being Mindful of Dynamic Populations

While considering how and what we teach, colleges and universities may bring an emphasis on equity to bear. Recent history has energized academics about the ways these institutions have or continue to marginalize, oppress, or actively harm campus populations. Maltreatment and structural inequity by gender, race, religion, class, sexual orientation, caste, geography, and other aspects of human identity and social participation has recently risen to the top of many institutions' agendas. It is not enough to focus a transformed teaching mission on what and how we teach, but also whom we educate.

There are now many calls for addressing educational inequities. Justice has, after all, been a theme of pedagogical discussion since Socrates. More recent advocates have included Paolo Freire, Myles Horton, and bell hooks. In terms of teaching climate change, there are some examples. One from New Zealand exemplifies the idea, within a democratically assembled document addressed to that nation's leadership: "Education and training developed by Māori for Māori will be important for reducing existing inequities, as research indicates that current education and training providers are not serving Māori well."[33]

Another population is involved in academia's teaching practices. Alumni sometimes have a personal stake in how their former institution teaches. This can take the form of connections to beloved faculty members or campus events. Alumni sometimes return to campus to give inspirational or practical advice about their careers. As the climate crisis continues and deepens, current students might scrutinize those alumni for those career choices. How will students surviving desertification think about alumni who work at energy companies? Will current students ask an administration not to accept charitable gifts from former students who went on to employment with legal firms that defend Big Oil, served with politicians who fought climate mitigation laws, or worked in financial houses that kept coal mines going against the tide of public opinion? Alumni working for Aramco, Chevron, Gazprom, ExxonMobil,

National Iranian Oil, BP, Royal Dutch Shell, Coal India, Pemex, Petróleos de Venezuela, or PetroChina may once have been sources of pride and emulation. How much longer will that attitude persist? Alumni relations may become more difficult.

Academic institutions worldwide may also expand their teaching of another, more at risk population. Climate refugees are likely to grow massively in number as the Anthropocene proceeds. They will cross subnational and national boundaries in search of safety and better lives. Each area subjected to the worst pressure should see its population move to less devastating climates. As people exit the ocean-threatened lands around the Bay of Bengal and expanding aridification and desertification in Spain and western China, they will have many needs.

Colleges and universities will then face the question of whether to aid them. This question falls out in two parts: giving physical assistance, such as temporary housing, food, and clothing, and providing postsecondary education, either in person or online. For some, the latter may seem out of place—surely the more important question is physical safety? After all, colleges and universities may offer to support those material needs, depending on their situation and capacity. Academic institutions can provide shelter, food, medicine, and clothing, as well as serving as evacuation guides and interfaces with governments. Yet climate migrants will need or want higher education for a range of options. First, if they travel to a new linguistic area, they may seek language instruction.[34] Second, if they cross into a new legal jurisdiction, they may seek training on local professional requirements for their field. Third, if they cannot or choose not to perform their previous jobs, they can study for a new field, as do many non-refugee adults. Fourth, refugees may wish to study about their new homes, or about the climate crisis, or whatever their passion or curiosity requests—again, like many non-refugee students.

We are already seeing early signs of this. Brian Fagan writes of one nation pedagogically supporting the migrants of another: "The Australian government is supporting the training of small numbers of young

Kiribatan people in employable skills like nursing, on the assumption that they will remain in Australia instead of returning to their overpopulated, threatened homeland. The president of Kiribati supports this program, which he calls 'migrating with dignity.'"

Some forward-looking activists have called for the definition of migration as a positive, not negative, strategy. Colette Pichon Battle asks us to think of "migration as a public benefit." Sarah Stillman urges us to recognize the deep, varied, and complex lives of migrants, then to see them as signs of global change, akin to the monarch butterfly. Roughly 80 million people are already involuntary migrants worldwide, by one count.[35]

We might view academic support for climate refugees as a kind act of charity or instead as a calculated attempt to develop a workforce that could accept lower compensation. We could ask which academic fields are taught to migrants and which are not. Grappling with such efforts, and offering them, is likely something that those within and adjacent to academia will do more of as the twenty-first century progresses. Should a given college or university teach climate refugees? How should this be done in a humane and sustainable way? Should such instruction integrate with material support? Is this something that can be done with private resources or state ones? Such questions are likely to occur with increasing frequency.

Some academics may see the audience for their teaching in even broader terms, as the public at large. Already, there are researchers who have turned to various media and platforms to share their knowledge and expertise far beyond the confines of a classroom. For example, in Canada's British Columbia, the Adaptation Learning Network, a university-led network, teaches online classes about climate topics to anyone in that province—or, in practice, anywhere in the world. In Upstate New York, the State University of New York College at Oneonta runs a public education project that invites participants to collaboratively create a post-carbon future.[36]

Summing up, as universities and colleges wade into the climate crisis, they may rethink their teaching function. Curricula, pedagogy, and taught populations can shift. Learning about and reacting to the Anthropocene in its full expression may constitute a new form of liberal education. We may see new uses of technology and a rethinking of how to support students. Ultimately, academia could revise its understanding of how teaching relates to the broader community. That campus-community relationship is the subject of our next chapter.

II THE WORLD, THE COLLEGE, AND THE GLOBAL EMERGENCY

CHAPTER 4

The Transformation of Town and Gown

What is being fought for is an identity, a culture, a beloved place that people are determined to pass on to their grandchildren, and that their ancestors may have paid for with great sacrifice.

—NAOMI KLEIN

At a practical level, what would it mean to say, I acknowledge this land, and I'd like to learn my responsibilities so that I can become a proper guest? . . . What does it mean to be an educational leader on Kumeyaay land? What does it mean to be a provost on Indigenous land? What are my responsibilities?

—JACK TCHEN, ERICA KOHL-ARENAS, ERIC HARTMAN, AND K. WAYNE YANG

When we think of colleges and universities, we often think of them in isolation. We may imagine individual researchers, students, or classes doing their work. We may also envision the physical plant of a campus, as we did in chapter 1, its imposing buildings and, sometimes, attractive grounds. Yet all too often, we neglect the extent to which an academic institution is intertwined with its local community and ecosystem. American academics gave us the fine phrase "town-gown relations," where "gown" refers to the ceremonial gear we sport for graduations and "town" stands in for whichever community abuts the academic space (urban sector, rural area, or an actual town). Town-gown relations is the subject of this chapter, both in the United States and around the world, as climate change impacts everyone without much regard for our human boundaries. A heat wave, firestorm, or flood threatens a campus along with its neighbors. And those human boundaries, which historically can be either fruitful or contentious, or both, may become even more so on both counts as the Anthropocene unfolds.

We can start exploring this topic by considering the town (again, also an urban area or rural zone, depending on circumstances) as a physical presence. Our exercise from early in chapter 1 is worth repeating, at least as one for the reader. If we think of the many physical components of a human settlement adjacent to an academic space, we swiftly catalog nearly the entire built environment that civilization presently offers. Office buildings, roads, airports, farms, power plants, sewage systems, hospitals, homes, abandoned factories, parks, military bases, prisons: the point here is to realize just how extensive a so-called town's presence is in the world, and thereby to grasp how exposed all of that is to various climate pressures. Again, the reader is invited to consider the many ways those pressures can play out, from advancing deserts to rising temperatures, surging waters, warped ecosystems, droughts, and so on.

This gives us a mental model with two major components, the campus and the town, each facing the Anthropocene. They are connected along many axes, as a moment's reflection will show. Some faculty, staff, and students live in town and off campus. In terms of economics, some academics also work in town, while some who live in town decide to take classes at that institution. Students, faculty, and staff spend money in town at restaurants, rental car agencies, tattoo parlors, rental apartments, and so forth, contributing to the local economy. The campus may purchase food, electrical power, construction materials, and other goods and services from local vendors. In terms of identity, we may view a campus as playing a crucial role in how we understand a community, as with Oxford and its university. The reverse can also be true, as we think of a given academic institution's identity as bound up with its surrounding community shaping that nature strongly, as with Georgetown University and Washington, DC. What we value in one may include what we value in the other. Naturally, this relationship can involve politics in a number of ways, from one protesting the other to regulation and elections.

To a large extent, the "town" will confront the climate crisis for the next century alongside the "gown." Growing desertification can silt up city streets and campus walks alike. Rising seas break over boundaries both municipal and academic. When its source glaciers evaporate and the Ganges River shrinks as a result, or even stops flowing, riparian communities of all kinds will be injured. Both academics and their adjacent communities now confront climate change together and are forced into strategic thinking and, hopefully, foresight.

Those communities do act independently of their campus, of course. Much as we've already discussed universities doing, a municipality or rural district may consider climate policies, practices, and projects of all sorts, from redesigning buildings to re-sourcing electrical power, building walls against rising tides, adding cooling shelters, and more. We can pick some examples from the United States, whose federal structure and contentious politics provide a rich variety of case studies. A county authority could ban certain aspects of the carbon economy from functioning, such as oil drilling or refining, as one did in Washington State and another attempted in Maryland. Cities and towns could pass measures to reduce, modify, or ban residential and business lawns as poor uses of fertile ground that could be put to better, more sustainable uses. Local authorities can also create or expand post-carbon transportation spaces, including bike paths and lanes or routes for foot traffic. The desire for more walkable cities seems likely to persist. Communities can grow or electrify public transit, or do both. They may add microgrids to their reliance on larger power grids for greater local control and flexibility.[1]

Larger-scale efforts can reshape those academically adjacent communities as the area around a campus struggles with other climate mitigation issues. Various instances of the carbon ecosystem may stand out for regulation, opposition, critique, or even violence, from oil wells and refineries to fossil fuel company offices. Built environment upgrades and replacements are likely, depending on resources. For example, the City of Ithaca in Upstate New York recently launched a massive plan to overhaul its buildings for the crisis: "The focus [is] on retrofitting buildings—installing electric heating systems, solar panels and battery storage as well as reducing energy use and greening the electric grid." The cost for this is estimated at $100 million, raised entirely from private capital. At a larger scale, one Massachusetts city published an ambitious plan to redesign its waterfront against rising sea levels:

> In East Boston, a deployable floodwall system has been designed across the East Boston Greenway, while Constitution Beach will be redesigned to combine flood protection with recreation and access. Key transport corridors such as Bennington Street and Main Street will be elevated, with $4.8 million in funding already committed.
>
> A series of parks including Christopher Columbus Park will be elevated to protect against flooding, as will the Harborwalk.[2]

Farther south, we can consider the interesting case of Miami-Dade County, an area adjacent to and about the level of the Atlantic Ocean. The United States Army Corps of Engineers provided a climate defense plan to the county, which turned it down, citing expense and blocking some charming views of the water. The Corps offered a second, revised plan, which, while much less expensive, is not as thorough as the first. It will leave some areas vulnerable to some flooding and damage, more so than in the initial proposal.[3]

There are nearly twenty campuses within ten miles of the shore, and a cluster are within a city block of the water's edge. Not only does each institution have to consider its physical vulnerability, as discussed in chapter 1, but it must now factor in the contours of this new and lesser plan. To what extent will the Corps' structures protect their respective physical plants? How will their off-campus populations be impacted? The climate change decisions of local authorities can have huge effects on higher education.

Local governments may also deem methane emissions from agriculture, food waste, natural gas, and other sources as a problem worth addressing. Government mandates can drive local producers to capture or burn off methane produced in coal mining and oil drilling. Businesses, governments, and nonprofits may experiment with methane digesters. The success or failure of such efforts can impact colleges and universities

Select colleges and universities near the Atlantic Ocean, in Miami-Dade County. Image created with Google Maps

as an instance of the tertiary climate crisis force mentioned in the introduction to this book. Academics can also study methane control attempts, as a group of scholars recently did in examining China's regulations of methane emitted from coal mining.[4]

Gown Meets Town

Co-creation of solutions through partnerships and engagement of relevant stakeholders including academia, industry, government is critical in research and innovations—the "Triple Helix." What is less emphasized is the role of communities and how they ensure relevance, acceptance, and learning from community experiences—the "Quadruple Helix."

—RHODA WANYENZE AND BARNABAS NAWANGWE,
MAKERERE UNIVERSITY, UGANDA[5]

Why does this matter for a community's academic neighbors? To begin with, a town's climate transformation may serve as an object of study for researchers in various fields. Faculty in urban studies, civil engineering, political science, and history may find documenting and analyzing these changes to be fruitful. Their students, both graduate and undergrad, may also participate in this research, doing fieldwork while developing their disciplinary skills. Many academic departments can follow this path, as we've seen in previous chapters.

Additionally, a campus may learn from its neighbors. A city or town actually building and maintaining a local seawall nearby can teach practical lessons about local geology for a university considering one of its own. Further, campuses can also benefit from the climate actions of their neighbors. A community that protects its wetlands, peatlands, or forests affords thereby a range of gifts to its academic denizens. Lastly, nearby climate actions may inspire an academic community to follow suit or attempt similar projects. A college's board of trustees that initially decided not to alter its historical buildings might revisit that choice once the local town amends its historical buildings with solar panels and upgraded insulation, simply through the force of a real-world, tangible example. A campus near a local alternative energy project might see its faculty, staff, and students inspired to participate, compete, and collaborate. For example, Atlanta's The Ray project, which develops roads that generate electrical power, stands near a group of colleges and universities, including Georgia Tech.[6]

A community that doesn't take steps to respond to the climate crisis can teach other lessons. The political defeat of a pro-climate cause can show an administration how to better support its own members. Conversely, a university leadership opposed to conducting climate policies may find useful tactics in a neighboring climate denial settlement. We should expect NIMBY (Not In My Back Yard) attitudes among local governments as well, a stance that might nominally support climate actions yet dislike them taking place on site.[7]

As noted earlier, the local community and an academic institution are not ultimately that separate, being interconnected in many ways. We can now think more closely about the relationship between local government, local politics, and universities, starting with how the town may seek to influence its gowned neighbor. For example, a local government may seek to influence postsecondary curricula, especially for public institutions. Such pressure could involve increasing or banning climate change material or shaping it in particular directions. The State of California has already considered such a law, albeit for primary and secondary school. This would require climate change to be "a coursework requirement for students in grades 1 through 6, and a graduation requirement for students in grades 7 through 12."[8] We could imagine communities encouraging or directing learning about specific mitigation approaches, local ways of knowing, religious views, or connections to specific industries.

A local community can issue laws, make policies, or otherwise act in ways intended to alter an academic institution's behavior. A county judge ordered the University of California, Berkeley, to freeze enrollment growth due to a complaint about rising student numbers having a deleterious environmental effect on the community. We can question the sincerity of such complaints (Were the litigants just using ecological law as a cover for old-fashioned NIMBYism?) while also recognizing that such legal mechanisms are in place and can be used with campuses. Such authorities could also implement laws and policies intended to protect specific populations within a community from climate problems, with which campuses must comply. A recent paper modeling mortality rates for people subject to unusually hot and cold weather found seniors (aged sixty-one and older) to suffer the worst; it is not implausible to expect city and county authorities to allocate resources to protect those populations. Colleges and universities may be pressed into service, such as making campus cooling shelters available to off-campus elders. Local authorities could also use the threat of elder deaths as a discursive tool to persuade academic institutions to take more climate mitigation steps in general. A countywide mandate preferring electric cars may compel a campus to speed up its vehicular decarbonization. A city ban on the use of electricity from fossil fuel sources may drive a university to switch to renewables that much more quickly. How might a campus and community work together to provide water access during a severe, persistent drought?[9]

Such rules may also concern climate refugees, their movements, legal rights, and support. The twenty-first century has seen a series of these, ranging from total exclusion and transportation to other sites to setting up quotas, housing, and job support. Climate refugees are likely to grow to very large numbers, which can yield difficult discussions in the town-gown relationship. Think of a divided approach, when a campus offers to host refugees while a city or county prohibits this, or the reverse. Imagine a city assigning a group of migrants to a college, or academics protesting a town's pro- or anti-refugee stance. This assumes refugees come from elsewhere. If there is a local disaster that renders locals homeless, then academic-community relations become even more challenging. What is the responsibility of a college to host nonacademic populations in the case of a flood or fire?

And what is the opposite duty for a nearby area, when parts or all of a campus are wrecked? The 2020 Lightning Complex fire in California gave a preview of this when it rendered some Cabrillo College students, faculty, staff, and even a trustee temporarily homeless. The college responded by carefully surveying its population, shifting some instruction online, making a large parking lot available for vehicles as ad hoc shelters, donating food and water, and setting up gym space for indoor safety.[10] How many local communities are preparing to address such crises now?

Enforcing local laws and policies will involve police forces, either locally situated or drawn from elsewhere, including a nation's military. Police may also react to on-campus political movements around climate change, from spontaneous protests to long-planned acts of violence. The long history of vexed relationships between police forces and campus populations seems likely to extend into the rest of the twenty-first century. We could imagine investigators probing a university population and eliciting resistance, or students, staff, and faculty protesting police they deem to be unjust for climate and other reasons.

Economics

Changes to local economic conditions can exert various levels of economic pressure on a university or college. This much is obvious from even a casual analysis, as we see when a local economic boom presents more options to an academic community, while also pricing some academics out of goods and services through gentrification. The opposite is also true, as local recessions reduce community options while also making the area less desirable for potential students, staff, and faculty. Both may occur in many locations because of climate-driven dislocations. Swiss Re forecasts a massive financial blow to the world by midcentury unless humanity manages to mitigate the crisis significantly: "Global temperature rises will negatively impact GDP in all regions by mid-century." Even if all nations succeed in controlling global temperature rise to below 2 degrees Celsius, the world could suffer a gross domestic product drop of more than 4 percent. In the Swiss Re forecast, a 2 degree temperature rise causes an 11.0 percent GDP drop; 2.6 sees a 13.9 percent fall, while 3.2 generates an 18.1 percent collapse.

This will not fall out uniformly. Such hits are distributed unevenly. As a leading reinsurance firm observed, "Economies in south and southeast Asia are particularly vulnerable to adverse effects of climate change, and advanced economies in the northern hemisphere least so." That uneven distribution should play out within nations and down to localities. We have already seen instances where climate disaster strikes a community and its higher education institutions are forced to consider their actions. For example, as of this writing, Madagascar is now in its fourth year of climate-driven drought, which may cause starvation in the populace. What is the role of the University of Madagascar in such a situation?[11]

One way such economic woe can occur is through a successful and significant decarbonization effect. Industries depending on fossil fuels may collapse and unemployment soar unless the transition to a new economic base proceeds effectively by developing sufficient successor enterprises. Academic implications follow directly. Think, for example, of campuses in regions with substantial coal mining, plastics manufacturing (based on petroleum inputs), air travel hubs, or businesses that produce oil-burning devices. Their reduction could seriously injure the community around an academic institution, minus a humane transition. Consider nations like Iraq whose economies depend heavily on producing and selling oil. Absent developing an economic replacement, communities within petrostates will suffer economically, and academic institutions will feel the pain as well. Rising crime around a campus, increasing substance abuse, and net emigration to other regions do not bode well for the health of a college or university.[12] Moreover, uncertainty around these dynamics could drive away investors seeking more stable vehicles, again hitting communities financially. Put another way, a college or university that successfully positions itself to survive the climate crisis internally can be economically injured by the decline of its neighbors.

Local taxes and climate change add another economic dimension to the topic. To begin with, institutions that depend significantly on local public financing are, of course, closely bound to the community's economic health. They could face severe austerity if the locality reacts in a cutting way to a climate-driven downturn. Additionally, local authorities facing economic pressure will sometimes be tempted to increase taxation on and around a local college or university: raising charges on property, for example, or asking for more support from the campus. We can foresee another taxation option, as municipalities and counties try to build up funds for climate change efforts. One way of doing so has already appeared in California and involves taxing the highest energy users. The idea is to either disincentivize those users from draining so much power or to boost revenue from their undiminished behavior. This might place academic institutions in a bind: pay more in certain energy taxes and support the community, or use less energy while benefiting from others consuming more?[13]

Culture

An academic institution can contribute to shaping a local community's culture and vice versa. One way this could appear in the climate context is through what some call green or climate gentrification. This is the process whereby investment in climate mitigation mechanisms—seawalls, flood barriers, barriers against sand—attracts more real estate financing and ultimately boosts prices. Poorer and marginal communities are pushed away. This can accelerate when private finance, as opposed to public monies, takes the lead. Worse, if climate mitigation efforts are spatially limited—covering one community but not its neighbors—populations may be forced into less well-protected situations. Would a college's or university's climate efforts cause this effect? Or would another campus be subjected to nearby green gentrification?[14]

Postsecondary institutions can play another role in shaping local culture, in terms of gender. Academic institutions educate increasingly large numbers of women. The proportion of university students identifying as women worldwide rose from under 10 percent in the 1970s to over 41 percent as of 2020, according to World Bank data. To the extent that those women are from or, if from elsewhere, remain in the local area, this process can gradually transform patriarchal societies and sexist attitudes by giving women more options and social mobility, interacting with other social and political movements. This may also benefit communities seeking to anticipate and mitigate the climate crisis, as women are generally more likely to spend more money on environmental issues, as well as to follow more eco-friendly behaviors. Women are also more likely than men to be hurt by climate disasters, so preparing them for what may occur is especially important. Educating and supporting women is a powerful contribution from an academic institution to its nearby community.[15]

Collaboration

In addition to cultural, economic, regulatory, and research connections, town and gown can also meet in productive collaboration. Each can bring to bear its strengths and capacities as they seek to cope with the climate crisis.

Some collaborations rely on academic research abilities. Rutgers University worked with several New Jersey entities, including the Jacques Cousteau National Estuarine Research Reserve, to produce a detailed report on how best to protect local, vulnerable populations from climate damage. In New York State, Colgate University joined the neighboring town and village in 1998 to conduct and share research about preparing for climate challenges. Results have included presentations, discussions, grants, land restoration, and plans for the future.[16]

Other collaborations blend research, teaching, and community action. Dickinson College's Center for Sustainability Education began including the local area through community research and service learning, then expanded when students worked with the town's borough council to measure municipal greenhouse gas emissions. This was successful enough to inspire the council to ask students to craft an action plan, as well as for the local county to request similar services.[17]

A Rhode Island offshore project connected local businesses, governments, primary and secondary schools, and higher education. WindWinRI situated turbines in high-speed air corridors, bringing generated power onshore. It also provided educational experiences for younger students, in addition to job training and curricular development opportunities for undergraduates. The University of Rhode Island has hosted an Environmental Fellows program "to engage and train students passionate about a career in sustainable energy."[18]

Some projects undertaken by the local community may appeal to members of the academic population. For example, a local nonprofit might launch a biochar creation center; some students might ask their institution to participate and set up biochar sites on campus. A local religious group could launch a climate initiative, which a religious studies professor decides to research. A clinic appears, focusing on climate crisis–created diseases, either of the body or the mind. This attracts interest from college faculty and staff. These are opportunities for productive town-gown interactions.

Some activism may bring food systems into the town-gown matrix. A campus movement might choose an off-campus site for agricultural revival, deciding to clear land and rebuild its soil. A local project to encourage small-scale or community gardens for food production and insect pollination might inspire academics to join in or to create their own. An academic experiment with growing food plants amid a forest could inspire nonacademics to attempt their own. Campus activists might urge off-campus farmers to adopt no-till planting.

On the infrastructure level, local collaboration could take many forms. If either a university or its nearby community creates its own electrical power, it might sell some of that to the other, either for money or as an exchange of goods or services. Town and gown might share electrical storage units for locally produced power. The community as a whole, academics included, may proceed along a path of shared electrical development and exploration over years. Areas where water scarcity is a growing problem may see joint projects to reduce water loss.

We may see campuses and communities interacting on a grassroots level through climate activism. Students may organize a movement to decarbonize their academic world and find common cause with similar activists in town. Faculty and staff interested in lobbying politicians around global warming issues could ally with members of the adjacent county. Marches and demonstrations may move between campus and off-campus spaces. We have already seen examples where campus populations or units lobbied local authorities to block or alter development, as when the College of William and Mary helped stop a major homebuilding project aimed at a coastal wetland.[19]

Such combined activism can involve clashing political drives on a campus, obviously. If members of a college or university participated in such a demonstration, tensions could rise with colleagues who opposed what the march called for. Academics are, famously, more than capable of arguing over different points of view. These divides can interact with nonacademic party politics. For example, the 2018 Sunrise Movement protest of US congressional leader Nancy Pelosi represented a split within the Democratic Party. We should expect academics and their neighbors to hold and express opposed opinions based on their affiliation with these broader political divides. In short, regional or national climate politics can echo across campuses and their adjacent communities. This dynamic can flow in other directions, as when Boston College students, frustrated by trustees' resistance to their divestment campaign, took their cause to the Massachusetts attorney general in an effort to outflank campus obstacles. In other words, campus conflicts can use external authorities for leverage.[20]

On the other hand, some communities may resent campuses for their successful climate adaptation. This could take the form of ideology, with climate deniers and anti-greens resenting what they see as an institution's politics, or irritation at the perceived wealth and clout a campus possesses, especially in areas with growing income and wealth inequality. The experience of economic or climate disaster runs the risk of enflaming town-gown tensions as well.

At this point in our discussion, we will move on to discussing higher education's public engagement at a larger scale, that of subnational and national governments. But before we proceed, one more point about local communities remains. People can invest a great deal in a physical site, be it campus grounds, a town, a rural tract, or part of a city. Historically, we often find meaning in the relationships between ourselves, land, local history, and how we see these interact. We can invest the connections with psychological power, seeing a campus as the place where our careers took off or we met a life partner. The town may be where a child was born, a project was completed, or a terrifying event was survived. Socially, there are many ways to think about and feel these connections. A tie to a town or campus may link us to a group or community, from alumni to sports fans, fellow religious sectarians, or professionals. This is one source for the psychological pain we have discussed earlier, such as solastalgia.

As we move more deeply into the climate crisis, we may bring a new or recovered ethos of care to such spaces. Some climate activists describe the struggle they engage in as one of restoration, restoring a site to a better state and time before fossil fuels warped the world. This may take the form of post- or anticolonial politics, as with one of the quotations with which we began this chapter: "I acknowledge this land, and I'd like to learn my responsibilities so that I can become a proper guest... What does it mean to be an educational leader on Kumeyaay land? What does it mean to be a provost on Indigenous land? What are my responsibilities?" There is a sense of treating the earth better than we have done since the Industrial Revolution began, of respecting Indigenous ways of relating to the natural world. Along these lines, the Land Relationships Super Collective refers to its work as "a network of land and water-based projects. It is a contingent collaboration of autonomous efforts to decolonize and heal relationships to land."[21] Others view academic and academia-adjacent zones as terrain teeming with potential innovation and progress. My point here is not to valorize or commend one of these views, but rather to remind us that postsecondary institutions and their neighbors are, among other things, places we can invest with a great deal of meaning and possibility. We should keep these thoughts and sentiments in mind, alongside the other academia–climate crisis connections we've outlined so far, as we envision how they may develop through the rest of this century.

CHAPTER 5

Academia in the World

The future is about old people, in big cities, afraid of the sky. —BRUCE STERLING

Academic institutions are deeply involved in their local communities, as we saw in the preceding chapter. They are also engaged with large-scale entities, from provincial or state authorities to national governments, corporations, and nonprofits. As the world progresses through climate change, a range of actors and forces will impact colleges and universities as they grapple with the global crisis. Put another way, higher education is and will be subjected to state policies, social movements, and civilizational currents in general. Academics may also step forward on regional, national, and international stages proactively.

Exploring the many intersections between the academy and the world is the purpose of this chapter. As with other chapters, this one draws on evidence from recent history in order to build models of possible futures.

Provincial and State Authorities

We can start by scaling up our analysis from local cities and counties to consider the impact of subnational provincial or state actions and policies. Public direction of higher education can occur at the national or subnational (i.e., state, provincial, departmental) level. It can shift from one to the other over time, depending on changing politics. Provinces and states can issue their own regulations that impact academics. For example, the Canadian province of Quebec recently decided to financially support consumers trading in gas-powered vehicles for electric cars. Interstate water distribution agreements offer other examples, especially those whose policies, institutions, and expectations have built up over decades, as in the American West. The COVID-19 experience offers some useful parallels, as when more than a dozen American states banned public (state-supported) universities from implementing certain pandemic policies, such as requiring tests or wearing face masks. States or provinces can also weigh in on the vexed question of cyberinfrastructure's carbon footprint. The State of California—home of Silicon Valley—issued regulations in 2016 that sought to restrict certain computer hardware designs, such as monitors and graphics cards, that drew higher amounts of electrical power than other parts.[1]

Provinces and states can, of course, also take steps that do not mitigate but instead exacerbate the climate crisis. The state government of Texas recently spent one-third of an emissions *reduction* fund to . . . widen highways, which will most likely *increase* driving and hence boost emissions.[2]

National Governments

Next, we can consider the many ways nation-states can interact with academic institutions in light of the climate crisis. Obviously, there are variations in space, as each nation can present different relationships to its postsecondary sector than others. There should also be variations in time as political and bureaucratic orders succeed each other over the twenty-first century, a succession partly shaped by how a polity perceives the climate crisis's impact. What we consider here is a range of possibilities; not every campus will experience, or even have access to, all of them.

Certain government agencies may become crucial players in helping communities survive the Anthropocene, including academic ones. Funding support is a major function. For example, the British government has set aside funds for forest and watercourse improvements, including forest floor protection. Should campuses lobby to have access to such grants, and in so doing compete with other entities, both academic and otherwise? The obverse of funding climate mitigation and adaptation efforts is governments taxing GHG offenders, as we have seen in Indonesia. That nation's finance minister, Sri Mulyani Indrawati, argued for such a measure both to mitigate emissions and to generate more state revenue. One academic question on this score is whether public institutions should support such policies, both because of their climate ends as well as in the hope of gaining more funding for education.[3]

Governments can provide other resources of interest. Most create and publish both data and information about climate topics, while some even make available an infrastructure for their content, as with NASA in the United States. In copyright terms, many national governments' published content can enter the public domain directly, making it more available than materials behind paywalls. As the climate crisis unfolds, some, even most, governments may expand their publication operations, giving academics more resources to work with. Of course, some may cut back or warp their data and information, depending on local politics, which presents colleges and universities with challenges. On a different level, we should bear in mind the possibility that national governments could orchestrate upgrades to their regional and national power grids, supporting multidirectional electricity flows, which allow for local generation and sharing of power. Naturally, such changes impact physical campuses.[4]

National governments can alter the political or physical environment of a college or university as part of an unrelated policy. Support for methane- or carbon-intensive industries can make it more difficult for academics to take public or even institutional steps against emissions. State support for direct air capture, as with Britain's recent decision to open four facilities, may make it easier for universities to mount DAC projects. Similarly, the British decision to plant more trees and protect peatlands could provide a supportive context for colleges seeking to do the same. The United States Department of Energy organizes the Wind for Schools program, which connects primary and secondary schools with local wind power generation sites. Regional state universities serve as facilitators in this effort. A French policy reducing jet travel in general may frustrate academics who cannot reach research or collaboration destinations easily through rail.[5]

National governments can also issue policies requiring campuses to take or avoid certain steps as part of broader policies. Mandatory reporting of climate data and operations may prove widespread, as states seek to both understand and direct greenhouse gas emissions. We saw early signs of this in 2021 when Britain's chancellor of the exchequer asked G7 nations to "impose mandatory reporting of environmental risks on their big companies . . . Under the proposals, the biggest companies would report annually on their exposure to risks and opportunities presented by climate change." The US Securities and Exchange Commission (SEC) published expanded rules for businesses to report on greenhouse gases emitted by suppliers and other partners.[6] Colleges and universities are generally not the size of major firms, but they can be at risk as well as able to take innovative steps, as we've already seen. Imagine such reporting policies expanding in coverage to cover the largest businesses, then moderately sized ones, and so on down the scale. Nonprofits, such as most of academia, would not be immune. We could see a campus chief operating officer or chief climate officer working throughout each year to prepare that annual report, and then the institutional leadership reacting to the government's reception of it.

Once some form of reporting or other climate assessment is in place, governmental pressure to reduce emissions is a logical next step. This could take a variety of forms, from executive directive to legislative mandate or judicial order. State pressures should also vary in strictness and compulsion, from mandates requiring immediate action to polite encouragement through funding or appeals to civic duty. It may involve political ideologies, as in China, where the Ministry of Ecology and Environment recently established a Xi Jinping ecological civilization thought research center to integrate that Marxist ideology into the climate crisis:

Xi Jinping's thought of ecological civilization is an important part of Xi Jinping's thoughts on socialism with Chinese characteristics in the new era. It is a major theoretical achievement created by General Secretary Xi Jinping based on the practice of ecological civilization construction in the new era. It is a scientific guide and powerful for the construction of socialist ecological civilization. Ideological weapons are rich in connotation and far-reaching significance. To study and explain Xi Jinping's thoughts on ecological civilization is a great responsibility and a glorious mission.[7]

Governments can also crack down on certain technologies, as when the Chinese state banned bitcoin mining and usage. This entails conformity from campus IT policies as well as of all potential technology users affiliated with an academic institution. Put another way, national climate policies can change academia's computing environments.

Preparing such greenhouse gas reduction policies may involve well-known national political resources, actors, and levers of change, including lobbying, lawsuits, fundraising, endorsing politicians, media campaigns, backroom deals, and less savory methods. The point is that the state can seek to compel colleges and universities located in its nation to change their behavior, often as part of a broader policy direction. For example, South Korea's leading party asked the nation to cut greenhouse gas emissions by 40 percent in under a decade, representing an enormous transition for every level of society. South Korean universities do not appear to be exempt from this drive.[8]

The Chinese government is already ordering all public institutions, academic and otherwise, to reduce greenhouse gas emissions over a multi-year period. Xinhua stated that "by 2025, the total energy consumption by public institutions will be kept within 189 million tonnes of standard coal." Water usage was also subjected to this directive: "Total water consumption will be kept within 12.4 billion cubic meters." Beijing's National Development and Reform Commission is taking steps to crack down on high-GHG projects, as it "recently urged local governments to reduce more than 350 planned projects, and pledged to take down those that have failed to meet government requirements." Doing so at scale will likely entail expanded surveillance of energy use as well as government sanctions: "While some regions approved high energy-consumption projects illegally, the NDRC noted there are other industries and companies that set overly ambitious goals, while some financial institutions had cut credit lines for coal power projects. The state planner said it is working with relevant departments to formulate work plans and to ensure 'no deviation' from central government policy."[9]

We could imagine governments cracking down on digital communications in order to slow the spread of climate denial information and organization. This could occur in various ways, depending on the nation, such as the state encouraging digital businesses to quash digital content it deems counter to its climate strategy, having internet service providers block certain terms and people, or outright censoring such material. Popular dislike of some internet-carried and platform-accelerated content could encourage governments to do this. Furthermore, the technological infrastructure could change in line with this impulse. Again, this could take many forms, such as incorporating climate data and user behavior modification into Internet of Things devices. Imagine a smart refrigerator monitoring household red meat consumption and locking users out once they exceed their weekly allotment, or a car that prohibits drivers from driving more than a certain number of kilometers per month. We could see either software or hardware implementing content controls when users create text, images, sound, video, and so forth, prohibiting or cautioning them from uploading materials that firms deem counter to climate mitigation or understanding the climate crisis.

Given such current and emerging practices, we can foresee more ambitious or detailed national regulations addressing the many dimensions of the climate crisis as it impacts organizations: the kinds of vehicles owned, building standards, capital investment sources, sourcing of electrical power, food served, travel, and so on. Governments can also address themselves directly to academia, targeting our sector's particular features, including supported research, curricular content, and cocurricular activities.

Climate-related changes within governmental structures may also encourage academic responses. Campus leaders may view new policies and operations for state bureaucracies and militaries as precursors to policies that will be aimed in their direction in the future. Individual faculty members and administrative leaders may also seek to impress governments (as funding sources) by following suit with state actions, or simply because the policies sound like good ideas. A transformation of government buildings, for example, can have a significant impact on emissions if the government is large enough and the changes ambitious enough. Shortly after taking office, the Biden administration directed federal properties under renovation or construction to follow Energy Star standards. How many campuses will take that under advisement for their own architectural work? In Britain, the University of Reading launched a program to develop new climate curricula for primary and secondary schools nationwide; how many university faculty and staff will consider expanding their own in response? Universities can also act to anticipate government actions on other fronts that impact their climate crisis work. For example, Texas A&M University shut down a collaborative climate science lab with Chinese partners, in part because administrators reacted to a growing federal crackdown on allegedly malign Chinese interventions within American academia. "The lab was on a list of foreign agreements that the [Texas A&M University] system wanted the College Station campus to terminate because of national-security fears."[10]

The real picture will most likely be more complicated than that. Not all academics will want to embrace certain responses to climate change, as we've noted elsewhere, and could urge quiet opposition, open dissent, or simple defiance of governmental actions. Faculty, staff, and students will debate one another on such topics, based on the historical evidence. At the same time, governments can also pursue courses of action that do not mitigate the climate crisis. Indeed, some will actively worsen conditions through expanding carbon burning or decreasing forest size. As with their political opponents, such administrations could impact higher education through general measures, targeted policies, or both. To make things even more complicated, the same national government can make policies that simultaneously mitigate and exacerbate the climate crisis. We have seen this with Britain's Boris Johnson government, which seeks to become a global green leader while also trying to burn more coal. One part of a government can drive changes in another's climate position, as when a German court's ruling caused that nation's cabinet to alter carbon emission goals, or when the Dutch supreme court ordered that country's government to follow a similar path.[11]

Government actions may specifically target higher education. The previously mentioned Philippine law requiring students to plant trees offers an early and small example of how this may occur.[12] We can envision other possibilities. A state may tie funding to institutional behavior, such as rewarding or punishing campus climate adaptation work. A government could order one or more campuses to provide climate relief services to certain populations, depending on the moment and the politics. A state could also encourage, mandate, discourage, or ban the use of certain climate practices and technologies, from installing geothermal energy on-site to decreasing food waste. Regulations may attend the interaction of higher education with arms of the state, such as the military or elder services. Imagine campuses tasked with housing military units on climate relief missions, or providing climate education to pensioners.

At the same time, we must remember that while national governments may act to mitigate or adapt to the changing climate, they may also exercise their powers to suppress climate actions of all kinds. In a Newtonian sense, every effort to address the climate crisis may be met by an equal and opposite state response. We saw one very small sign of this in 2020 when a British antiterrorism unit listed Extinction Rebellion as a dangerous extremist group. The panoply of domestic monitoring and control options available to the modern state can all attack a wide range of climate activists.[13]

In short, academic institutions face a varied governmental landscape at the national level. As in all cases of national directives and encouragement, we confront the question: How should academia respond? Academia's options may be as varied as the governmental terrain they navigate. For example, if a national government mandates overall GHG cuts, what is the role of academic research? The state can use scholarly articles to support this policy. Does research or teaching that does not concur with that policy risk defying the nation-state? Should individual researchers and instructors, not to mention academic administrators and support staff, monitor and consider publishing scholarship that supports or opposes government policy? Should instructors nudge student discussions toward the national agenda and gently quash the opposite? The degree to which academic institutions and individual academics see themselves as autonomous against state direction affords an important opening for action. Campuses, as well as students, staff, and faculty members, have the existential ability for independent thought and action. This could range from a single student protesting a government policy to multiple colleges and universities organizing climate action in the face of state opposition. As ever, academic freedom runs into contingencies and risks.

Universities and colleges can act toward governments on another level. Over the next eighty years, we should expect to see academic activism targeting governments and policies. Students and faculty already play roles in Extinction Rebellion and Greenpeace. In 2021, Norwegian protesters occupied that nation's energy ministry and temporarily blockaded a major oil shipping port. Others sought to block parts of other world cities. How many of them had ties to academia? As the Anthropocene progresses, such activity could escalate to broad dissent, acts of violence, or open rebellion, depending on the institution and its situation.[14]

International Connections and Multinational Organization

We can now scale up our discussion of government relations to the level of multistate and international politics. Thinking of the impact of national governments on how higher education responds to the climate crisis, we can start by including multinational public institutions. Some of these are locally based, such as the Mexico–United States water rights organization, the International Boundary and Water Commission, which wrangles a truly complex set of demands and interests.[15] Other water organizations will come under pressure as droughts, floods, and storms ratchet up. Academics may find partnering with them, or at least understanding and researching them, to be productive.

Nations consisting entirely or mostly of low-lying, small islands are often on the leading edge of climate catastrophe and can organize to fight for their interests, including their survival. In 1990, thirty-nine of these nations formed a global organization, the clearly named Alliance of Small Island States, to advocate for themselves collectively on the world stage. Members are drawn from the Caribbean, the Indian Ocean, the Pacific, and the Atlantic.[16]

For its part, the informal organization of the G7 nations has significant global influence due to the group's combined wealth, military might, and geopolitical reach. It recently considered imposing carbon reporting policies on its countries, a policy that could, by itself, speed along GHG reductions while inspiring other nations to follow suit. On the other hand, tensions may unfold between the G7, whose members generate a slowly shrinking proportion of global carbon output, and the rest of the G20, whose share and total amount grow. Some non-G nations may also oppose the group's actions and stances for anticolonial reasons as well as ideological differences from the right and left.[17]

Larger international groups may play larger roles. The European Union has a great deal of potential power in the climate world, given the clout and wealth of its member states. It has taken some steps already, including organizing the creation of carbon-sequestering carbon sinks across that continent and its Parliament passing carbon laws to apply to all participating nations:

> EU forests, grasslands, croplands and wetlands altogether removed a net 263 million tonnes of CO_2 equivalent (CO_2e) from the atmosphere in 2018, according to the European Commission. That tally also accounts for the amount of CO_2 released when trees were cut down or wildlands burned.
>
> The Commission will next week propose a target to expand the EU's sink to absorb 310 million CO_2e per year by 2030 by giving each member state a legally binding goal, according to the draft.

The EU's "Fit for 55" plan pushes to increase carbon prices and so encourage other sources of energy. A European Commission proposal aims to force airlines to cut down their carbon emissions by discouraging them from using current fuels. The EU also contributes billions of euros to developing nations most likely to be damaged by the climate crisis.[18]

The North Atlantic Treaty Organization (NATO) overlaps with the EU greatly yet has some different members (notably the United States) and a separate focus (military security). Accordingly, NATO can approach climate change as a matter of joint security, which can then lead to a range of possible efforts. The alliance could become a major disaster relief provider, seeing natural calamities as destabilizing events. It might plan for the impact of climate forces on its members, such as drought afflicting Mediterranean coastal countries. Externally, NATO may see its mission including protecting its members from other states that climate change has made more dangerous, or to block high numbers of climate refugees from arriving.

The African Union also has a great deal of potential for advancing climate work on that continent. Its Great Green Wall of the Sahara and the Sahel (Grande Muraille Verte pour le Sahara et le Sahel) is a multinational success story. Millions of hectares of land and tens of millions of trees now retard the Sahara's expansion. This is also a very long-term project, having been supported by the AU starting in 2007 and only being about 15 percent accomplished so far.[19]

At the biggest scale in world international organizations, the United Nations has a potential leadership role in this area, at least in terms of convening discussions and also by allocating some established resources. Historically, the UN has played a powerful organizational role in some climate actions. As Tufts University professor Abiodun Williams observes:

> The UN has advanced our understanding of climate change through the assessment reports of the intergovernmental panel on climate change created by the UN Environment Programme and the World Meteorological Organ-

ization in 1988. The UN has also facilitated the creation of the climate regime based on three international treaties: the 1992 UN framework convention on climate change, the 1997 Kyoto protocol, and the 2015 Paris agreement.

We can add the series of Conference of the Parties (COP) meetings to that list. UN leadership can encourage or criticize national decisions, as when secretary-general António Guterres charged G7 nations with failing to live up to promises made to the developing world. The United Nations High Commissioner for Refugees is already addressing some climate refugees and is thereby in a good position to do more as that problem expands. The UN Educational, Scientific and Cultural Organization (UNESCO) is also playing a role in the refugee world, developing an improved qualification passport for people escaping one country for another. That new document could support climate migrants.[20]

All of these organizations present many faces and possibilities to higher education. Academia can certainly research them as they develop, and that scholarship may become important in decision-making and public debates. Postsecondary institutions can similarly teach this topic. Political science and international relations are the obvious fields, but other disciplines have historically explored international action, from history to economics. The climate dimension brings in a range of natural and social sciences as well. On a more material side, international institutions offer potential relationships for colleges and universities, some of which they have already realized: internships, career placement, and grants. Other relationships may appear as the climate crisis deepens. If a national government may regulate higher education, it may do so as part of a multinational, collective action. A multinational resource agreement, such as cutting back water sharing or incentivizing the use of certain materials in construction, would impact campuses directly.

We should also bear in mind that multinational collective action can proceed without being grounded in established global institutions. History is replete with examples of ad hoc alignments and alliances that proceed without crystallizing into anything more formal. Treaties can suffice for action, as the great Montreal Protocol on Substances that Deplete the Ozone Layer (1987) did to prevent that part of the atmosphere from dwindling. The climate crisis can easily drive the creation of more of these. For example, the problem of advancing electrical power generation across Southeast Asia involves multiple nations, especially in considering damming rivers flowing between them. Scholars have argued for the benefits of a unified system. The 1999 Energy Charter Treaty, initially created to integrate former Soviet Bloc nations' energy enterprises with those of western Europe, still regulates a great deal of the world's international energy exchanges and trade. It specifically includes rules for improving efficiency and lessening environmental impacts, although the success of the latter can be debated. More recently, India and Britain agreed to climate collaboration in a new trade deal, including London investing US$1 billion in Indian green projects, along with an effort to raise private capital for similar efforts. At the same time, India linked its GHG reduction strategy to a need for more foreign investment. Costa Rica and Denmark launched a Beyond Oil and Gas Alliance with the goal of "manag[ing] the decline of oil and gas production." Just before the COP21 meeting, China announced it would end investing in other nations' coal plants and would instead consider applying those funds to green energy projects abroad. Such relationships can, of course, be determined by historical relationships of empire, the Cold War, and other dynamics.[21]

Such relationships may arise out of international tensions along climate lines. For example, a carbon tax in one nation can mean higher prices in another if the latter imports products from the former. The second country will have a financial interest in working with the first to reduce that passed-on price, or to come up with some other means for addressing it. Australia offers an interesting example of this: it has avoided imposing its own carbon tax but is considering paying a tariff for outside products instead. We saw another version of this in 2021 when the Chinese government, already committed to reducing carbon dioxide emissions, nevertheless protested a European Union's cross-border carbon tax as bad for economic development. Similarly, projects in one nation have significant environmental impacts on neighbors, as when a Polish coal mine drained water from an aquifer shared with a nearby Czech town, leading to European Court of Justice battles between the two countries' central governments.[22]

Of course, rather than developing an agreement, a nation could instead erect tariffs or trade barriers against other nations based on the former's assessment of the latter's environmental damage, as the Australian case suggests. Such trade, policy, and ideology clashes could well escalate to the formation of larger alliances, economic warfare, and actual kinetic fighting. Obviously, such conflicts have implications for the full spectrum of academia's work, from research to community relations and campus operations.

On the other hand, there may be more calls for more international cooperation and organization as the Anthropocene deepens, especially as climate issues cross human-crafted borders at will. Geoengineering projects, too, can cause impacts across continents and seas. Creating them should involve multiple national partners. Should damaged parties sue geoengineers in The Hague, or will a new global tribunal address climate crimes? Perhaps competing arbitration platforms will appear, or a transnational climate authority will emerge.

A national government can declare a climate emergency, as the Philippines did in 2020. This can, in theory, give the state more powers for action, both domestically and internationally.[23]

Depending on the nation and its perception of the crisis, how the state uses those powers can range from nothing to the symbolic to totalitarianism. Powers for taxation, property seizure, surveillance, policing, regulation, and so forth are all available to governments looking to mitigate or adapt to the climate crisis.

Taken together, we can now see how individual academic institutions must see themselves interacting with multiple layers of government, potentially from transnational entities down to local communities. This can take the form of dealing with competing political divides. How does a college or university navigate a division that spans multiple governmental layers? A province might align with an institution's climate work, yet the local neighborhood and also the national administration pursue opposing policies. Over time, these positions can switch, with the adjacent community becoming more friendly while the next governmental layer up turns hostile. There are many examples from the present day, such as when China's National Development and Reform Commission strongly criticized provincial authorities for failing to attain climate goals. The State of Victoria set its own emissions plan against that promulgated by Australia's central government. Another case involves the American state of Louisiana, where gas and oil companies have long played a major role. A recent proposed law sought to declare the state a "fossil fuel sanctuary," aimed at blocking national laws. It would have "ban[ned] local and state employees from enforcing federal laws and regulations that negatively impact petrochemical companies." Do academic institutions try to avoid or resist such state laws, or do they follow them? There is also the open split between the International Energy Agency (IEA), which called for an outright ban on investing in additional fossil fuel extraction, and the governments of Australia, Japan, and the Philippines, which argued that there were other ways to reduce carbon dioxide emissions, including "future negative emission technologies and offsets from outside the energy sector." Campuses may have to navigate these conflicting governmental positions, which can be enormously difficult, especially as climate pressures ratchet up.[24]

We should also remember that climate policies and support will differ by nation, sometimes greatly. Different populations as well as administrations hold varying understandings of and attitudes toward the climate crisis. Examples are easy to find. Support for more coal mining drops in some areas but can be found in the Polish state, which extended the life of one coal mine. Similarly, some national leadership denies climate change is a problem, as with Brazil's Jair Bolsonaro government, while China created a new national group of climate leaders. These are snapshots of policies, caught in a moment of time. Governments change policies and administrations, of course, as do political parties. National governments can also contradict and reverse themselves, as we have seen with Britain's Johnson administration, which manages to support decarbonization and expanding coal mining at the same time.

Similarly, World Bank economist Branko Milanović observes that Norway presently acts in a deeply contradictory way. On the one hand, it relies heavily on petroleum extraction for exports, which generate very lucrative rewards domestically; on the other, it is formally committed to decarbonization and electric cars. In fact, Norway's pro-carbon economy literally and financially fuels its domestic move toward a post-carbon order.[25]

Colleges and universities in each nation have to calibrate their climate actions in response, especially for their own national governments.

Impact of Corporations

In addition to governments, other world entities present higher education with challenges and options. Corporations obviously play a major role in the global economy and thereby impact societies in many ways. One reason to consider them at this point in our discussion is to recall our earlier point about the ways local and regional economic changes, driven or shaped by the climate crisis, can hit colleges and universities. Businesses mediate, cause, or participate in these transformations; their actions and status are in a sense the face of economic change. Postsecondary institutions experience recessions, booms, and industries crashing or taking off through these enormous corporate lives.

This touches in a very general way how businesses react to climate change. Corporations also make decisions around climate that impact academia. We can explore some examples from Australia. When a major bank stops financing coal, as Macquarie did in that nation, a local instance of that industry could decline or fail, with economic and social fallout that falls on an institution. A major investment enterprise in one nation can pressure another country to change its energy investments, with multiple ripple effects in the latter, some of which touch postsecondary education. We saw an example of this when Dutch investment firm Robeco Institutional Asset Management pressured Australia to shift away from fossil fuels. Investors can call for climate action from any entity seeking funds, with potentially enormous effects. To return to the example of Macquarie, that financial services business used recent fossil fuel profits "to launch a revamped climate policy that includes exiting coal by 2024 and using its multibillion-dollar clout to pressure heavy emitters into accelerating plans to reach net zero targets." Businesses can also compete on the green front, seeking to outdo one another in climate achievements. The case of two Gujarati billionaires, Mukesh Ambani and Gautam Adani, competing with each other on this score may point to a form of competition likely to appear more frequently as the climate emergency continues.

Those are secondary or indirect effects; we can easily see more direct forms. Funders can insist on including or avoiding certain features in a campus capital project, such as a carbon-positive building. Investors may request more transparency on financial or climate matters from campuses, as some already pressure other enterprises.[26]

Academia may take a more proactive role with the business community. Campuses could partner with technology companies in pursuit of green computing. Tech firms can also provide climate services that campuses may desire, such as Google providing Google Map results that show routes requiring lower carbon emissions. Academics can also protest or even attack companies for their perceived role in creating the climate crisis, of course, and this may become the more noticeable attitude. Colleges and universities can additionally identify companies engaging in greenwashing in order to stop their institutions from contracting with them.

Civil Society

Academia also engages the world through nonprofits in many forms. Students, staff, or faculty may work with nonprofits locally or abroad, and I can see no reason for such efforts to avoid the climate change domain.

To begin with, nonprofits can play an important role in climate mitigation. For example, the Mission Innovation nonprofit, funded by Bill

Gates, focuses on supporting new forms of energy, including addressing the problem of wind power's variability. Organizations like 350.org (which began in a college!), C40 Cities, Project Drawdown, GenderCC, and the Climate Action Network can offer global information, consultation, internships, practical experience, and career options to colleges and universities. Regional nongovernmental organizations (NGOs), like the Caribbean Community Climate Change Centre and the Arab Forum for Environment and Development, can fulfill similar functions. A number of nonprofits are mission focused on education about climate, like Action for the Climate Emergency and the Climate Change Education Partnership Alliance. Academic partnerships with that kind of nonprofit are easy to envision.[27]

NGOs may attempt to influence campus operations, either directly or indirectly. An example of indirect influence may be discerned when Greenpeace criticized China for burning coal to power growing numbers of data centers and 5G base stations. To the extent that academic institutions rely on those technologies, the nonprofit is implicitly hailing them for their dependence or complicity. After all, independent nonprofits working in the policy space hope to influence state actions, civil society, and public opinion. Think tanks play a role in policy making, and their work can have direct or downstream impacts on higher ed. Governments can engage think tanks and nonprofits to support their positions or yield intelligence, based on their having some measure of independence. For example, the British Climate Change Committee, hired to assess the United Kingdom's climate efforts, issued a scathing report, finding the efforts too little and incompatible with support for fossil fuels.[28]

Religions, some of the oldest nonprofits, can also play a potentially strong role in the climate crisis. As noted in chapter 2, religious scholars are already researching connections between the faiths they study and the Anthropocene. Such work can be done within the academy or can occur in the broader public sphere. For example, Texas Tech University professor Katharine Hayhoe, both a climate scientist and a Christian activist, publicly seeks to connect the two roles. At an institutional level, religions can seek to influence colleges and universities on multiple levels, from working with affiliated institutions to pressing for changes to public higher education.

Academic institutions working with nonprofits need to conduct due diligence to make sure that efforts are not wasted or worse. For example, the Nature Conservancy offered a carbon offset service to clients, using fees to protect forests. Yet a Bloomberg Green investigation found at least some of that effort misdirected, spurring an internal investigation of that service.

Ultimately, higher education can work with a mix of partners, including nonprofits, businesses, and governments, depending on the situation. A fascinating example is that of an Aarhus University wind turbine project, which saw that Danish academic institution working with two businesses and a nonprofit think tank.[29]

Major Social Changes and Trends

Beyond relations with civil society, businesses, and governments at the national scale, academic institutions engage, and are impacted by, cultural and social trends concerning climate change. The unfolding crisis will influence and shape those forces while creating new ones. How academia reacts to the overall crisis will be partially mediated through such trends.

To pursue this point requires a great deal of caution and even humility. The farther we extend our vision into the future, the more numerous its possibilities become. A quick glance at any given eighty-year period in human history shows our capacity for social and cultural creation. We can point to major forces already in operation at the present moment—globalization, what some call the fourth industrial revolution, progress in human rights across numerous domains, antiracism, religious conservativism, and neonationalism, to name a few—and imagine various forms they can take over the next generation, even to the next. Their interactions are complex and numerous, yielding a deep matrix of possibilities. In addition, we can envision trends not currently discussed much at present, which might rear up to become world forces, such as a new religious movement, a political leader with the capacity of a Napoleon, or an ideology on the scale of Marxism. We might expect sea changes in public attitudes on key issues that we don't perceive as likely now. Just think how surprising the widespread contemporary drive for gender identity justice would have looked to forecasters in 2000, to pick one example. In addition, black swans of all kinds are always possible. Forecasting them precisely is famously challenging. The point here is that looking at broad currents in forthcoming history is a fraught task. It is one we will attempt here, again with all humility.

We can start by pointing to one of the most massive yet still surprising and underappreciated trends sweeping civilization now: the demographic transition, whereby people have fewer children and live longer. Let us take that trend further and integrate it with the climate crisis. One possible result is that significant parts of the world decide to further reduce the number of children they have. Imagine each child-bearing adult having one or fewer (on average) children during the course of their life. The generally accepted number of children for maintaining a population's size is 2.1; below that, the group shrinks over time. Much of the developed world now sees rates closer to 1.6. Below that, societies will grow smaller at an even faster rate. If this occurs, will we value a smaller number of children more highly, leading to a new era in child-rearing? Will we see people or institutions shame parents for having children?[30]

For academia, will expectations for the standard of care practiced by colleges and universities heighten? How will institutions maintain their financial status if the incoming flow of students dwindles? For state-funded institutions, will governments decide to spend less as their schools teach smaller classes? For private institutions, how will they compete to win enough enrollment to keep operating? Alternatively, would younger students experience *less* social support than that enjoyed by previous generations as economic and environmental problems combine to elevate other priorities, or do states draft teens into various forms of climate action?

We could imagine such developments taking a differently gendered form. Naomi Klein asks us to think of fighting for climate justice in explicitly gendered terms. Not only is the struggle against patriarchy, in her view, but it is also a new "reproductive rights movement." She develops the point thusly: "What is emerging, in fact, is a new kind of reproductive rights movement, one fighting not only for the reproductive rights of women, but for the reproductive rights of the planet as a whole—for the decapitated mountains, the drowned valleys, the clear-cut forests, the fracked water tables, the strip-mined hillsides, the poisoned rivers, the 'cancer villages.'" All of life has the right to renew, regenerate, and heal itself.

This is not a call for encouraging people to have more children. Indeed, other thinkers, such as Donna Haraway, call on humanity to keep reducing that number and to develop new forms of family, including with other members of the animal kingdom. This combination of reducing human reproduction while encouraging nonhuman fecundity could appeal for a variety of reasons. If a significant number of people come to hold something like this view, we could imagine calls for colleges and universities to not only increase their climate work in general but also to more actively connect humanity with the nonhuman, natural world. Supporters of this view may also celebrate higher education's role in the education of women, as that historically reduces childbirth rates. There may be more expectations for academic institutions to support family planning. The opposite may also be likely, that opponents of this view blame postsecondary education for supporting such an agenda. This last will have implications for college and university reputation and financing.[31]

Financial changes can also occur in other ways, deeper and broader. As the Anthropocene continues, humanity may not maintain its current neoliberal political economy. If alternatives appear, how does academia respond? For example, we can imagine new arrangements that cut back on overall economic activity on multiple fronts, such as reducing purchasing and lowering the hours in a workweek, in order to cut emissions. A four-day workweek might produce fewer greenhouse gases, as Platform London recommends.[32]

Other plans and models are under discussion now, as we started to see in chapter 2. A circular economy would pause total economic growth in order to focus on redistributing the present settlement of goods and services in a more equitable way. This may entail shifting funds to secure the poor, the retired, and those economically marginalized by various identities, including race, gender, ethnicity, and religion, as well as relative vulnerability to climate damage. Such an arrangement would suspect the creation of new GHG emissions while also diverting resources toward climate mitigation and adaptation. Michel Héry and Marc Malenfer describe four possible ways Europe might embrace this paradigm:

1. Global corporations drive the circular economy to consumer approval.
2. Different parts of Europe engage in different timelines, although eventually all end up on the same page.
3. Global corporations turn Europe into a service supplier for growing middle classes elsewhere, failing to attain the circular economy.
4. Different European populations adopt circularity unevenly, and ultimate fail to cohere.

The differences opening up in that scenario quartet point to fault lines and possibilities within one continent, involving some of the actors and sectors discussed in this chapter so far. Obviously, multiple divergent paths are possible as well. These possibilities open up before academics trying to forecast how their societies might respond to the climate crisis—and just on this one point of macroeconomic order.[33]

An alternative political-economic model is that of degrowth, in which civilization actively reduces not only economic growth but its entire output, again in order to reduce greenhouse gas emissions, if more aggressively. Author and activist George Monbiot offers a good sample of this view:

> Everywhere, governments seek to ramp up the economic load, talking of "unleashing our potential" and "supercharging our economy." Boris Johnson insists that "a global recovery from the pandemic must be rooted in green growth." But there is no such thing as green growth. Growth is wiping the green from the Earth.
>
> We have no hope of emerging from this full-spectrum crisis unless we dramatically reduce economic activity. Wealth must be distributed—a constrained world cannot afford the rich—but it must also be reduced. Sustaining our life-support systems means doing less of almost everything.

Part of the degrowth necessitates rethinking human society in noneconomic terms, or choosing new metrics other than GDP: in short, a revision of how we understand collective life. Barbara Muraca goes much further than this simple sketch, identifying a group of degrowth branches and concluding that degrowth opens up the political imagination into realizing even more.[34]

Engaging such a new economic order—should it start to take hold—gives academia numerous challenges and ways to engage them. Multiple college and university curricula can teach these changes, while researchers in many fields can explore their present impact, historical bases, and possible futures. Things become more difficult as institutions attempt to work with local communities as their economies are disrupted. They will also have to grapple with funding in a world that is no longer growing, or is contracting; this is a radical threat to many privatized higher education ecosystems. No-growth and degrowth economic models reveal to what extent academia is predicated on assumptions of persistent growth—arguably the mentality that powered the fossil fuel explosion. To what extent can universities and colleges as we know them survive such an economic revolution?

Anti-individualism

In such futures, our technological infrastructure may plateau or degrade. An end to growth or a new era of degrowth cuts against the grain of fiercely profit-driven technology companies. At the least, such new orders would drastically reduce some aspects of digital development, such as the rapid release of new hardware devices. Additionally, the unfolding climate crisis can further downgrade the digital world by means of direct damage. Rising waters and storm surges can injure or destroy data centers, transmission notes, mobile phone towers, and other key elements of the networked world. Indirect damage can also reduce technological capacity if supporting enterprises collapse due to climate reasons or if public funding drops through crises and losing out to other, more urgent priorities. Colleges and universities that depend heavily on IT for everything from enterprise resources to online teaching will confront serious problems. This is an area where research and innovation may prove critical.

We may also see some populations and institutions actively resist technology, which leads to additional constraints. Some climate activists have argued that human development of, and reliance on, modern technology has worsened the environmental crisis as well as the human experience. At times, this can be targeted at climate technology, such as geoengineering, as well as perceived attitudes and structures that underpin it, such as techno-solutionism. This can build on various recent schools of thought, such as opposition to Silicon Valley businesses for privacy violations or for encouraging the spread of climate misinformation. For example, Naomi Klein has criticized geoengineering as magical thinking, reflecting a dangerous attitude toward the world. Klein goes further and opposes spaceflight and space exploration as similarly counterproductive and dangerous:

> Before those pictures [of Earth seen from Apollo 11], environmentalism had mostly been intensely local—an earthy thing, not an Earth thing. It was Henry David Thoreau musing on the rows of white bush beans in the soil by Walden Pond . . . It was Rachel Carson down in the dirt with DDT-contaminated worms. It was vividly descriptive prose, naturalist sketches, and, eventually, documentary photography and film seeking to awaken and inspire love for specific creatures and places—and, by extension, for creatures and places like them all over the world.
>
> When environmentalism went into outer space, adopting the perspective of the omniscient outsider, things did start getting, as Vonnegut warned, awfully blurry. Because if you are perpetually looking down at the earth from above, rather than up from its roots and soil, it begins to make a certain kind of sense to shuffle around pollution sources and pollution sinks as if they were pieces on a planet-sized chessboard . . . And all the while, just as Vonnegut wanted, any acknowledgement of the people way down below the wispy clouds disappears.[35]

If such positions become more popularly held and influential, they might lead societies to reduce their use of, and support for, the digital world. Again, this has the implications for academia we just cited. In addition, academic fields based on modern technology—computer science, to pick the most obvious—might come under governmental scrutiny or be subject to popular criticism and opposition.

A broad antitechnology cultural movement could narrow in practice. Specific technologies may attract particular opprobrium, either from novelty or because of the perception that they drive problems or disasters. Atomic power has suffered this fate for two generations, often following spectacular failures. It's possible that geoengineering projects will follow suit. Direct air capture of atmospheric carbon dioxide might also become the target of significant dislike. Some now see it as a distraction from the mission of reducing emissions. Others go further, as Carroll Muffett, chief executive at the nonprofit Center for International Environmental Law, stated: "If on any reasonable examination of CCS [Carbon Capture and Storage], it costs massive amounts of money but doesn't actually reduce emissions in any meaningful way, and further entrenches fossil fuel infrastructure, the question is: In what way is that contributing to the solution as opposed to diverting time and energy and resources away from the solutions that will work?"

Rather than opposing categories of technology, such as types of hardware or software, future polities might use climate concerns to regulate types of technology-enabled content or behavior. An *Atlantic Monthly* article once charged internet porn with contributing to global warming. It is not difficult to imagine critics or authorities condemning, taxing, banning, or otherwise seeking to control various types of digital content.[36]

Policies and practices to control climate misinformation content could easily apply to other forms of information that people deem harmful. We have mentioned identifying the carbon footprint of certain technologies in use, such as bitcoin mining; other digital tools and internet-enabled content could be similarly labeled and restricted. Think of a conservative authority seeking to curb internet porn for the amount of bandwidth and storage tube sites consume, for example, or an authoritarian government blocking content it deems disloyal or dangerous based on its alleged GHG footprint. Such practices impact higher education on several levels. First, there is the possibility of authorities involving academic institutions in technology and content monitoring and controlling. Second, the principle of academic freedom can run up against planetary necessity. Third, social movements and political actors can draw on academic research to drive their digital campaigns.

A different yet related climate politics can also embroil academia. It is possible that some will call out for punishments to be levied on those who enabled global warming to occur. We can foresee charges laid against fossil fuel companies, governments with large GHG footprints, airlines, beef eaters, owners of private jets, and others. As climate damages mount and human costs accumulate, the desire to exact justice on responsible parties will most likely build up. It is not a stretch to imagine civil suits for damages or trials in The Hague for crimes against humanity. Fossil fuel firms' documented efforts to slow down climate awareness and fog public debate may appear liable or criminal. A nation could easily find the ability to go after an alleged malefactor for the deaths of hundreds of thousands or millions of people. These are legal and licit responses; we should not be surprised by popular acts of vengeance, from hacking and property damage to assassination. Those who see themselves as potentially vulnerable to any of these acts may seek to protect themselves through private security, isolation, buying off local authorities, or striking back at their assailants. Ultimately, we could see battles over climate culpability waged around the world.

What is the role of academia in such a scenario? We could see academic research becoming passive ammunition in these battles, deployed by combatants. Should colleges and universities play an active role in identifying and punishing climate criminals? Would institutions contribute legal firepower or become host to social movements demanding oil barons in chains? Academic creativity can lead to a variety of other responses, from defending accused parties to suggesting alternative responses, such as restorative justice. At a financial level, how would such a GHG struggle impact institutional endowments and donations?

Colleges and universities could play a different, more direct role in an era when the world seeks to punish the greenhouse gas culpable. Some may charge academia with aiding and abetting the climate crisis, and not without reason. Accusers could point to faculty who taught and researched petroleum engineering, which trained generations of fossil fuel producing workers and enabled further oil and gas exploitation. Critics could point to the small yet influential number of campuses that accepted funds from a variety of GHG-related sources, from agribusinesses that emitted methane to shipping firms and oil enterprises. They could also dun academics perceived as being too close to climate-denying politicians.

Politically, opponents can turn an academic strength against institutions by criticizing colleges and universities for contributing to the industrial economy that created global warming. Higher education's success in preparing graduates for careers played a key role in expanding the civilization that generated planetary emissions. To the extent that campuses taught students to consider the natural world an endless resource to be exploited, or not to consider the planet as an essential context in life, critics can charge postsecondary education with complicity in the global catastrophe. The legal charge of ecocide discussed in chapter 2 could become a serious weapon wielded against colleges and universities.

Moreover, if tempers run high enough—such as during or right after a spectacular climate disaster—critics could charge postsecondary institutions for simply not doing enough to address the crisis. In other words, while some academics may criticize and seek to change the world on climate matters, the reverse may also occur.

Academics at Large in the World

We can widen our analytical aperture even further and consider how academia approaches the rest of the human race, including not only states but whole populations. After all, there is an ancient tradition of faculty taking public stances through speeches, writing, presentation, consulting, media engagements, and more. The internet has amplified the range of what Americans call "public intellectuals" even further, adding many more platforms, from YouTube to blogs.

Already, professors participate in the world's climate crisis discussions. Some do public research that wins media and policy attention. For example, the World Weather Attribution (WWA) project, a collaboration involving European and North American universities and labs, tries to identify links between climate change and extreme weather events. Other faculty members publicly engage in pro-climate activism. We have seen this in many cases, such as Penn State University's Michael Mann, who brings his academic work to bear in public policy. Andreas Malm, a faculty member at Lund University, has participated in many demonstrations and some direct action, calling for, while practicing, climate militancy. Academics also intervene in public discussions to criticize climate actions. An Oxford University inorganic chemistry professor joined the Global Warming Policy Foundation, a very public climate crisis denial group. Several students have already joined climate lawsuits.[37]

We know from the historical record that academia can play some roles on the world stage, beyond the town-gown relationship. College and university research enters national and global discourse to some degree on its own. Beyond that, some academics have already played more active roles on the topic of climate change. Faculty have acted in the public intellectual capacity. One example saw Queensland University of Technology's Amanda Gearing actively documenting horrendous bush fires and sharing that knowledge, mediated through her academic abilities, with a world audience through The Conversation website and other venues. We have also seen academic faculty intervening in public discourse and decision-making, from scientists testifying before national legislatures to researchers publicly and collectively urging action on methane emissions. In chapter 2, we noted H. Tuba Özkan-Haller's recommendation that the academic tenure review process should recognize and support such interventions in the public sphere.[38]

Might academia play a larger role in the Anthropocene than it has done so far? In 2020, five British professors called on fellow scientists to agree to shared action and attitudes toward the climate crisis:

> We pledge to act in whatever ways we are able, in our lives and work, to prevent catastrophic climate disruption. To translate this pledge into a force for real change, we will:

- Explain honestly, clearly and without compromise, what scientific evidence tells us about the seriousness of the climate emergency.
- Not second guess what might seem politically or economically pragmatic when describing the scale and timeframe of action needed to deliver the 1.5C and 2C commitments, specified in the Paris climate agreement. And to speak out about what is not compatible with the commitments, or is likely to undermine them.
- To the best of our abilities, and mindful of the urgent need for systemic change, seek to align our own behaviour with the climate targets, and reduce our own personal carbon emissions to demonstrate the possibilities for change.

With courtesy and firmness, we will hold our professional associations, institutions and employers to these same standards, and invite our colleagues across the scientific community to sign, act on and share this pledge.

Note the combination of guiding personal actions, which might recall the Hippocratic Oath, and encouraging public communication.[39]

At a more intense or risky level than communication, might academics reach out to support climate refugees? Faculty, staff, and students could urge their institutions to offer such assistance in various ways (humanitarian, curricular, logistical) or provide such help on their own. Perhaps such calls at the present moment signal a forthcoming willingness for academics to focus energies on the crisis. It seems reasonable to imagine more such forms of postsecondary educational self-organization.[40]

To better anticipate such mobilization, we need more context. So far, we have been assuming a moderate Anthropocene, a middle path forward between extremes. Yet futures do not always stick to moderation, and we would blunt our forecasting imagination to think so. Now we can expand our sense of how the climate crisis may unfold by examining other scenarios: futures more benign and those much more dangerous.

III CHOICES FOR UNIVERSITIES IN A WORLD ON FIRE

CHAPTER 6

Best Case and Worst Case

Somewhere along the way, I had gone from being an ecologist to a coroner. I am no longer documenting life. I'm describing loss, decline, death. And that is what is accounting for my kind of overwhelming sense of grief.

—DIANA SIX, PROFESSOR OF FOREST ENTOMOLOGY / PATHOLOGY, UNIVERSITY OF MONTANA

The 58th COP meeting of the Paris Agreement signatories . . . concluded with a special supplementary two-day summing up of the previous decade and indeed the entire period of the Agreement's existence, which was looking more and more like a break point in the history of both humans and the Earth itself, the start of something new . . . The greatest turning point in human history, what some called the first big spark of planetary mind. The birth of a good Anthropocene.

—KIM STANLEY ROBINSON, *THE MINISTRY FOR THE FUTURE*

Thus far, we have proceeded in our futuring work based on a broad assumption about how the climate crisis may transpire. That scenario represents a rough middle path between extreme possibilities. It presumes some degree of human progress in handling greenhouse gas reduction as well as escalating climate pressures and disasters.

In this chapter, we consider two other scenarios, each of which embodies those other possibilities, those extreme futures. We will begin with optimism, outlining a path with the least climate damage foreseeable.

Next, we explore its opposite, a worst-case scenario. Neither is entirely precise, as each allows for substantial variation. We will see that imprecision in each narrative of how the scenario may come to pass, along with what that does to a changing world. Each discussion concludes with a model of what academia could resemble if their respective conditions bear out.

Please recall our hedges and qualifications from the introduction. These scenarios focus on a single trend, global warming's advance. That in itself contains a great deal of variables, ranging from geographic differences to probability estimates of major events occurring. They also try to account for an immense amount of civilizational variability, as we are, after all, trying to scry nearly eighty years of human history with the vast array of possibilities that entails.

Neither scenario is substantially shaped by what Nassim Nicholas Taleb calls "black swans," statistically unlikely events that, when they do occur, exert enormous impact. Historical examples include the 2008 financial crash, the September 11 attacks, and the emergence of some world religions. They are too large in number to admit serious consideration in this space. Black swan events will certainly adjust the contours of this chapter's scenario pair. I invite the reader to consider several and work out their potential impact, and hope for further research along these lines.

Table 1. Three scenarios for how climate change could develop through the year 2100

	Best case	Moderate scenario	Worst case
Temperature rise (degrees Celsius)	1.5–1.8	1.9–2.5	3.3–4.6
Number of major threshold events	0	0–1	2+
Sea level rise	2050: 1 foot 2100: 2 feet	2050: 2 feet 2100: 5–7 feet	2050: 4 feet 2100: 10–13 feet where records are still being kept
IPCC 2021 scenario(s)	SSP1-1.9	SSP1-2.6–SSP2-4.5	SSP3-7.0

Future emissions cause future additional warming, with total warming dominated by past and future CO_2 emissions

(a) Future annual emissions of CO_2 (left) and of a subset of key non-CO_2 drivers (right), across five illustrative scenarios

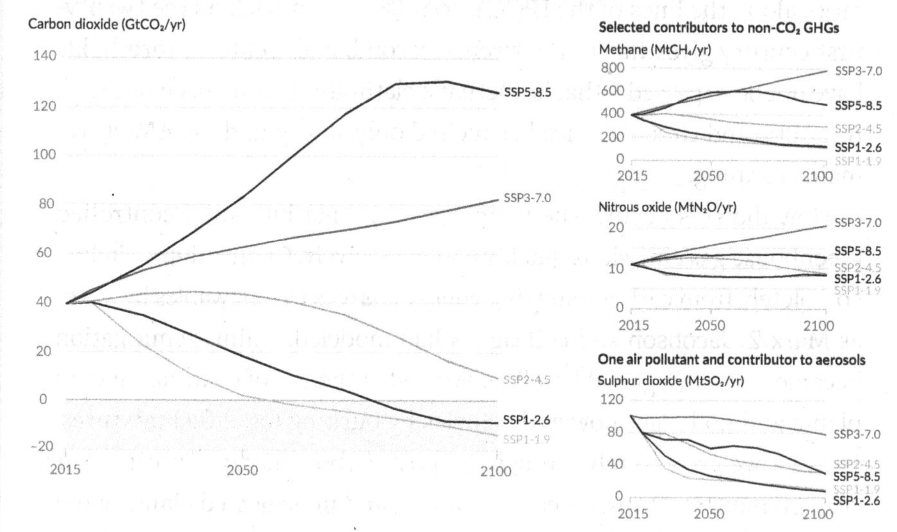

(b) Contribution to global surface temperature increase from different emissions, with a dominant role of CO_2 emissions

Change in global surface temperature in 2081–2100 relative to 1850–1900 (°C)

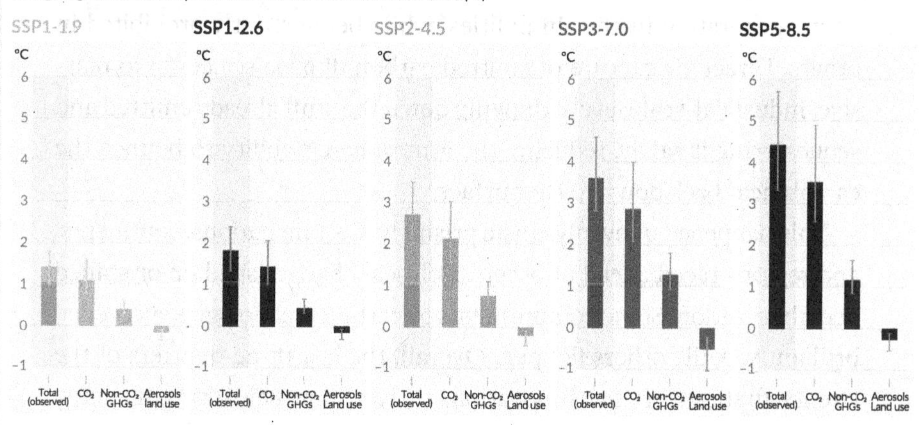

Total warming (observed warming to date in darker shade), warming from CO_2, warming from non-CO_2 GHGs and cooling from changes in aerosols and land use

Five IPCC scenarios for temperature rise during the last decades of the twenty-first century. Figure SPM.4, panel (b) from IPCC, "Summary for Policymakers," in *Climate Change 2021: The Physical Science Basis. Contribution of Working Group I to the Sixth Assessment Report of the Intergovernmental Panel on Climate Change*, ed. V. Masson-Delmotte et al. (Cambridge University Press)

Best Case

In this scenario, global temperatures rise 1.5 to 1.8 degrees Celsius by 2100, along the lines of the IPCC's 2021 SSP1-1.9 model. As the twenty-first century gives way to the twenty-second, major climate thresholds have not been passed—that is, methane clathrates haven't been released, Antarctic and Greenland ice has melted only slowly, and the AMOC remains running.

How did this occur? The main reason is that humanity controlled greenhouse gas emissions quickly and extensively. Civilization switched completely from carbon-emitting energy sources to renewables by 2050, as Mark Z. Jacobson and colleagues had modeled.[1] Climate mitigation became normal, a global baseline. We ended the use of coal, gas, and oil plants and no longer powered vehicles by burning fossil fuel mixtures. Instead, we grew to rely strongly on post-carbon energy sources: solar cells on most rooftops, on cars, on roads, and in dedicated plants; wind turbines, both on land and offshore; hydropower, using tidal motion and currents; geothermal. Nuclear power played an active role in planetary power generation in certain polities, while being strictly prohibited in others. Direct air capture of emitted carbon dioxide scaled up to massive, industrial-scale levels, drawing down the stuff already emitted and sequestering it safely, reducing the atmosphere's ability to bounce the earth's heat back down to the surface.

This happened unevenly, unsurprisingly. Certain nations, continents, and regions raced ahead of others. Political changes sped up or stalled decarbonization. Some commitments to certain technologies succeeded brilliantly, while others flopped. Overall, the last three-quarters of the twenty-first century saw innovations thrive. Some new forms of energy generation took off, such as various ways of generating power through hydrogen. Some power sources enjoyed a massive drop in cost, akin to what made solar power so attractive in the early twenty-first century. Post-carbon transportation systems grew over the decades, notably blimps and airships above and expanded rail lines on or near the ground. Airships for trips between cities rapidly cut flying's carbon emissions by

90 percent as people acclimated to slower speeds, assisted by onboard Wi-Fi.[2]

More innovations took place in science and human behavior. New ways to capture, store, break down, or reuse carbon dioxide proliferated. On the methane front, humanity rethought landfills in favor of extensive recycling and reuse. Our diets changed as well. By 2100, people tend to think of eating beef the way people a century before thought of smoking tobacco or dumping pollution in public spaces: as horrific offenses, dangerous and inhumane. Most people participate in growing food, either through personal or shared plots. We consume food produced nearby, mostly, avoiding that carbon-burning habit of the industrial epoch.[3]

Overall, we distributed both resources and costs more equitably than in centuries past—not without struggle, including several wars and revolutions, plus less photogenic but more peaceful conflicts waged through nonviolence and basic politics. Multiple new political economies have appeared. Economically, neoliberalism subsided gradually and was succeeded by several competing ideologies. Doughnut economies competed with more classically Green polities for global mindshare. Over the decades, the green economy yielded increasing benefits in terms of better quality of life for more people and more jobs, especially in the many levels of a transformed energy sector. Rapidly transitioning to renewables by the late 2040s saved the world trillions of dollars. Old-school oligarchies ruled some very unequal societies, while a greater number of others followed social democratic organization into redistributing goods and services more equitably. Various intellectual and cultural currents drove these shifts from the classic Washington Consensus: antimaterialist religious belief and spiritualities, anticolonialism, various forms of anticapitalism, Indigenous folkways positioning humans in balance with the rest of nature, potentially a successor to Xi Jinping thought. People sometimes refer to the twenty-first century as the Green era, as opposed to the preceding Gray or Brown century.[4]

After much experimentation and several small disasters, several geoengineering services are at work. Swarms of mirrors orbiting the earth tune the amount of solar radiation falling on our planet, monitored by a global utility and supported by an advanced, multi-planetary, international spacefaring ecosystem. The same body scrutinizes the world's oceans, adding several amendments to them to increase albedo or carbon capture. Massive schools of lawyers and bureaucrats are engaged in scrutinizing this utility's every move. The result: less global heating than anticipated.

The COVID-19 experience of 2019–2024 helped inspire humanity to take some of these steps. That pandemic gave us a glimpse of what cutting carbon emissions under our own willpower could look like, demonstrating that it was possible. Similarly, virus-driven reductions in air travel forced us to rethink the skies. Airlines never fully returned to pre-pandemic flight volume. The failure to organize effective countermeasures at an international level proved an embarrassment, leading to a series of interstate agreements over the following decades, culminating in a species-wide public health directorate.

Nonhuman forces also played a role in this scenario. Greenhouse gases already lofted into the atmosphere proved less powerful than once modeled. Natural forms of carbon sequestering proved more powerful. The earth's interior swallowed up more carbon dioxide than models had suggested.[5]

People developed more respect for nature, partly through rediscovering the fear of nature we once had. Even in a world mostly urban, there is simply more natural stuff present in daily life: more plants at work and home, more time spent helping produce food, more attention to animals. Rewilding has occurred in various forms, ranging from setting aside wildlife corridors and zones to bringing lost species back to life through DNA-based simulation, then reconstitution. Land use patterns have shifted by 2100, with scant acres devoted to raising cows and a bit less for raising crops, after the total human population declined. More land goes to host solar power, rock weathering, wind turbines, and DAC installations.

One natural wildcard was a major volcanic eruption in Iceland on a scale commensurate with Mount Pinatubo (1991) or Novarupta (1912). This form of nonhuman geoengineering cooled temperatures by ejecting huge amounts of particles into the atmosphere, which then circulated globally. This was anticipated by early IPCC reports:

> Based on paleoclimate and historical evidence, it is likely that at least one large explosive volcanic eruption would occur during the 21st century. Such an eruption would reduce global surface temperature and precipitation, especially over land, for one to three years, alter the global monsoon circulation, modify extreme precipitation and change many CIDs [climatic impact drivers] (medium confidence). If such an eruption occurs, this would therefore temporarily and partially mask human-caused climate change.[6]

The damage to lives and property in Iceland and elsewhere, first by lava and fires, later in the form of breathing and pulmonary disease, was either not discussed openly or else considered by those of a realpolitik disposition to be a painful yet fair price to pay.

One nonhuman factor created what some deemed to be false hope. Because solar output rises and declines cyclically, the midcentury downturn lessened the total radiation falling on the earth, giving some the impression that the crisis was not so dire as forecast. As the sun's output rose during the century's last decades, the rise in global temperatures confirmed the global green consensus, albeit with some scrambling and improvisation. By 2100, solar weather is a customary part of human weather news and discussion.[7]

Another fake best-case scenario is how some people viewed the unfolding crisis at certain times. When spectacular disasters were not occurring nearby or not shared widely through various media, people could convince themselves that all was well. On a related point, some failed to track gradual yet persistent changes over time as per the famous parable of a boiling frog—that is, without sudden and telegenic events, we can see things as roughly continuous and not disastrous.

While this is a best-case scenario, it is not a utopia. It is not all goodness and triumph. Despite the enormous transformation of human civilization, planetary temperatures still rise through 2050. As the 2021 IPCC report warns, "Global surface temperature will continue to increase until at least the mid-century under all emissions scenarios considered." This continued warming means a large amount of danger, with damaging events and harmful changes, including increased storm incidence and destructiveness. The most fragile ecosystems buckle. Humans will increasingly suffer from health problems. As one medical journal editorial observed in 2021,

> In the past 20 years, heat related mortality among people aged over 65 has increased by more than 50%. Higher temperatures have brought increased dehydration and renal function loss, dermatological malignancies, tropical infections, adverse mental health outcomes, pregnancy complications, allergies, and cardiovascular and pulmonary morbidity and mortality. Harms disproportionately affect the most vulnerable, including children, older populations, ethnic minorities, poorer communities, and those with underlying health problems.
>
> Global heating is also contributing to the decline in global yield potential for major crops, falling by 1.8–5.6% since 1981; this, together with the effects of extreme weather and soil depletion, is hampering efforts to reduce undernutrition. Thriving ecosystems are essential to human health, and the widespread destruction of nature, including habitats and species, is eroding water and food security and increasing the chance of pandemics.[8]

Urban areas are often even hotter than surrounding areas, due to the heat island effect, and both suffer from and cause extra problems. As the IPCC warned,

> Cities intensify human-induced warming locally, and further urbanization together with more frequent hot extremes will increase the severity of heatwaves (*very high confidence*). Urbanization also increases mean and heavy precipitation over and/or downwind of cities (*medium confidence*) and resulting runoff intensity (*high confidence*). In coastal cities, the combination of more frequent extreme sea level events (due to sea level rise and storm surge) and extreme rainfall/riverflow events will make flooding more probable (*high confidence*).

Sea levels rise as well, if not to the degree other scenarios project. According to the latest IPCC projections,

> It is *very likely* to *virtually certain* that regional mean relative sea level rise will continue throughout the 21st century, except in a few regions with substantial geologic land uplift rates. Approximately two-thirds of the global coastline has a projected regional relative sea level rise within ±20% of the global mean increase (*medium confidence*). Due to relative sea level rise, extreme sea level events that occurred once per century in the recent past are projected to occur at least annually at more than half of all tide gauge locations by 2100 (*high confidence*). Relative sea level rise contributes to increases in the frequency and severity of coastal flooding in low-lying areas and to coastal erosion along most sandy coasts (*high confidence*).[9]

All of the coastal problems we've discussed—flooding, health challenges, the options of defending areas or migrating from them—are still in play, if not as severely as they could be.

In this scenario, academia is more challenged than it was during 2000–2020. All of the challenges faced by higher education worldwide then are intensified by this least-horrible climate crisis. Improving quality of instruction, expanding research, and expanding access are major tasks made more difficult by rising temperatures, elevated sea levels, growing deserts, and all the social effects that follow. All of the American challenges of staggering student debt, declining public trust, plummeting public funding, under-supported students, and a casualized professoriate—to name a few—become more difficult as entities academic and otherwise cope with the unfolding Anthropocene. Colleges and universities may overhaul their physical campuses, transform their research enterprises, change their educational offerings, and alter how they interact with the local and global communities.

It is possible that resources will flow to institutions and programs that grapple with the crisis directly and effectively. We may see more faculty and staff jobs in climate issues, both academic and administrative, even as positions dwindle in the rest of the institution. This might lead to a kind of climate mission race, as departments and individuals argue for the Anthropocenic relevance of their work. Climate research and teaching may become central to some institutions.

Partnerships with the outside world predicated on climate issues in general may build goodwill and public support, insofar as people see academia as focused on the climate crisis. Academic institutions could portray themselves as trusted local allies, ready to bring their unique resources to bear. As communities in this scenario redesign themselves for the Green era, turning to colleges and universities will be good sense for many. Town-gown collaborations range from ecological projects to multiyear student work on building elevation, forest growth, and seawall expansion.

On the world stage, some academics will play major roles in informing and advising the world. National governments and international organizations tap researchers as policy advisors and communicators. Public intellectuals appear and become well known to large audiences. Academic content about climate change—in text, images, video, extended reality—circulates. Collaborations among academic institutions lead to the creation of several student climate organizations, which operate worldwide, combining service learning with research. Climate activism crosses the campus border steadily, linking the academy and the broader world.

Worst-Case Scenarios

We're fucked. —COMMONPLACE SAYING IN CLIMATE CRISIS CONVERSATION

Let us start with the bad news, then get to the truly terrible version. By 2100, global temperatures have risen between 3.3 and 5.7 degrees Celsius since the start of the industrial age, as per the IPCC's 2021 SSP5-8.5 model. Sea levels are three to five meters higher.

The natural world has been transformed. Oceans are more acidic than they were, which has killed off huge swathes of life, shattering ecosystems from local to intercontinental scales. Hopes for a "blue food" way of feeding humanity fail. Land-based species around the world migrate and/or perish, causing ecological ripples across food chains. Diseases spread within and between species.[10]

We realized this scenario through certain human actions. Our focus on short-term financial gain played a role, as we bound up much of society around quarterly profits for business. This vitiated long-term thinking and gave us plenty of reasons to keep emitting greenhouse gasses. As a result, we deployed alternative energy in only a marginal way for decades, relying on a mix of coal, oil, and gas as the anchor of our power needs. Nuclear power never recovered due to a mix of resistance based on perceived safety issues and concerns about investment and maintenance costs.

Human attitudes were decisive in making this future possible. Some can disable parts of society from taking or supporting steps to address the crisis. For one, a popular argument held that a temporary temperature rise above 2 degrees Celsius was worth allowing, so long as it enabled a drop in following decades. Yet in this scenario, the former occurred without the latter happening, a victory for short-term over long-term thinking. It is also possible that significant numbers of people acted through the twenty-first century as though the worst case would occur and gave up on both mitigating and adapting to the crisis, resigning themselves to catastrophe. Some speak of deep adaptation or terminal adjustments, citing the Dark Mountain movement as an inspirational foundation.[11]

Many responses to the COVID-19 pandemic pointed to this future. Despite some notable efforts, international collaboration repeatedly failed as individual nations took their separate paths: developing and marketing national vaccines, hoarding resources, blaming others. Humanity generally failed to come together to meet a common, global threat. In addition, some nations, such as the United States, devolved pandemic responses to the provincial and even local level, further distancing themselves from international action and policy coherence. Worse, the rise and success of anti-vaccine movements and disinformation campaigns demonstrated a deep human capacity for self-defeating and self-harm when confronting a complex, massive threat. One attitude embodied a kind of grim resignation, urging inaction against the virus because the world was simply dangerous and lethal.

As the century wore on, a similar attitude took hold and most people viewed global climate changes as normal and expected. As they acclimatized to each stage of the emergency, experiencing events as just par for the course, they became less likely to support extraordinary measures for what they see as an ordinary situation. David Roberts coined the term "shifting baseline syndrome" to describe responses to the COVID-19 pandemic over time, but it seems like a much older human tendency: to adapt to extremity and normalize it.

There is a scarier possibility, in many ways more plausible: we never really wake up at all.

No moment of reckoning arrives. The atmosphere becomes progressively more unstable, but it never does so fast enough, dramatically enough, to command the sustained attention of any particular generation of human beings. Instead, it is treated as rising background noise.

The youth climate movement continues agitating, some of the more progressive countries are roused to (inadequate) action, and eventually, all political parties are forced to at least acknowledge the problem—all outcomes that are foreseeable on our current trajectory—but the necessary global about-face never comes. We continue to take slow, inadequate steps to address the problem and suffer immeasurably as a result.[12]

Nature, of course, played the major role in realizing this worst-case scenario. The world crossed two or more threshold events that accelerated global transformation. One of these events could have been, for example, the AMOC current across the Atlantic Ocean slowing down significantly or, worse, stalling out. The northeastern shore of North America and northwestern Europe would suffer, perhaps counterintuitively, a sudden and serious temperature drop, with harsh consequences for local biomes, including food production systems. As one study modeling British agriculture concluded:

> Economic and land-use impacts of such a tipping point are likely to include widespread cessation of arable farming with losses of agricultural output that are an order of magnitude larger than the impacts of climate change without an AMOC collapse. The agricultural effects of AMOC collapse could be amelio-

rated by technological adaptations such as widespread irrigation, but the amount of water required and the costs appear to be prohibitive in this instance.[13]

Alternatively, the Amazon switched from being a carbon sink, capturing more greenhouse gases than it emitted, to becoming a net carbon and methane generator. This occurred when enough of the rainforests were destroyed and replaced by plains, especially for cattle grazing. The ice sheet on Greenland melted, starting from its western side, sloughing off vast amounts of water into the world ocean, which then rose accordingly. In Antarctica, Pine Island flows into the sea and the Thwaites Glacier gives way. Similarly, large amounts of methane have been released from clathrates in northern Siberia and North America, rapidly accelerating temperature rise. In this scenario, we assume two or three of these events have occurred or are under way.[14]

Moreover, each of these major tipping points or transformations can interact with others, accelerating temperature rise even further. The Arctic Ocean's switch from being covered with reflective ice to presenting only dark, absorptive water to solar radiation increased the ocean's temperature and sped its warming. The end of carbon sinks like the Amazon means more carbon dioxide is lofted into the atmosphere, further increasing temperatures. Increased heat and dryness in some areas mean a fire-vegetation feedback loop has kicked in, with lightning-caused fires lofting plant particles into the atmosphere and adding to heat retention, which dries out more foliage, creating more combustible material, while also increasing the number and intensity of storms, which makes more lighting possible.[15]

Humanity experiences extreme challenges, to understate the situation massively. Sea level rise drives many coastal cities to transform themselves through barriers and various forms of elevation. Some urban and rural populations migrate out from these threatened areas, depressing economic values and creating a vicious cycle heading toward collapse and the abandonment of cities. Beyond sea level rise will come increasing storm damage to many parts of the world, both in terms of frequency (more storms more often) and damage. Coastlines around the world have been subsumed, from India and Bangladesh to Brazil, Egypt to New York. Access to freshwater will become problematic as key water sources dry out (glaciers, snowpack), suffer from overuse (aquifers), or are contaminated by salt water. Storms will hit certain areas more often and with more forces, escalating in intensity over the years. Conversely, drought will afflict others, drying out jungles and forests while spreading deserts.

Food becomes harder to produce as a result of losing water, as well as from increasing temperatures. Some agriculture migrates to cooler lands, but it takes years to build up the necessary soil to support full production. Crops that do emerge provide lower amounts of nutrition. Burning plants and other materials adds more particulate matter to the air every decade, increasing breathing ailments and deaths. Ocean acidification drops the amount of edible fish humans can obtain. Food insecurity and sheer starvation advance. Diseases migrate and hit new populations; some mutate into new forms. The total number of deaths we can attribute to climate change are difficult to anticipate, but at least one source finds "the mean estimate of the total mortality burden of climate change is projected to be worth 85 death equivalents per 100,000, at the end of the century," or roughly seven million people. Economic devastation naturally occurs, costing at least 3 percent of global GDP. It does not strain credulity to expect not only social and cultural chaos but also political strife involving wars and collapse. Similarly, we may anticipate tyrannical states using advanced technology in ever more dire circumstances.[16]

One human response to this cascading disaster may well worsen it. We might attempt geoengineering projects that either fail or have massive, unintended consequences. National governments, corporations, supremely wealthy individuals, or other actors could well mount such projects on their own, with or without involving global oversight. Imagine one of the growing number of spacefaring nations establishing an orbital mirrors system. Software or hardware errors could lead to misdirected radiation on the Earth's surface, or successfully cooling one region triggers massive storms in adjacent areas. Consider a rogue billionaire who seeds an ocean with iron in order to build up carbon-devouring plankton, only to skew food chains and cause rippling die-offs among marine life, which then harm fisheries. An attempt to inject aerosols into the stratosphere succeeds in slowing temperature rises, until its funding runs out and the aerosols dwindle, leading to a rapid increase in global temperatures. As the situation worsens, nations may compete with one another to launch geoengineering projects, leading to, for example, a solar radiation management (SRM) arms race. In this worst-case scenario, we expect several such efforts, and disasters, to have occurred.[17]

How might academia respond to such a dire twenty-first century? The dangers to physical campuses outlined in chapter 1 become much more threatening here, with greater exposure to flooding, droughts, storms, desertification, and fire. A larger number of institutions will consider migration. Some of the academy's scholarly mission becomes ever more important as nations, companies, nonprofits, and individuals turn to it for desperately needed research and development. That mission falls under ever greater public scrutiny and suspicion. Some researchers and institutions are willing to take more risks than before, either with ethics or with dangerous technologies, from geoengineering to genetic modification, tailored diseases, and self-replicating machines. Similarly, some institutions ramp up their teaching of the climate crisis in order to graduate people who can best fight it or lead adaptation efforts. Campus-community partnerships become ever more urgent as both participants anticipate and suffer increasing stress.

On the world stage, some academic research scores attention or wins resources to attempt saving efforts. Other academics, perhaps tending to work in the humanities and social sciences, deem carbon-based civilization to be failing and seek to anticipate or prepare for a successor civilization. Younger people, including traditional-age undergraduates, see themselves creating that new world. Opposition from governments, nonprofits, religions, social movements, and nonstate actors may meet all of these academic enterprises at various points. Some will question academic freedom's value in an age of dire crisis.

That opposition can lead to cutting academic financial support. Other forces may reduce the academy's footprint as well. Some people who might otherwise have become students will respond to rising disasters by throwing themselves into working on them without postsecondary education. Social pressure might add to this impulse, with a popular argument being why waste time studying something that isn't the climate crisis, when the world is on fire? For private and privatized institutions, this can reduce enrollment and hence revenue, leading to cuts and shrinkage. As for public colleges and universities, governments could easily slash funding, citing the more urgent demands of fighting the climate catastrophe. Additionally, campuses may close on their own terms, citing the well-worn precedent of wartime. Overall, higher education could dwindle as this scenario approaches 2100, measured by number of institutions, enrolled students, or total staff.

But it could be worse. The preceding is not, in fact, the worst-case scenario. The Anthropocene has truly apocalyptic potential.

Let us take things further. Higher and higher global temperatures could trigger a runaway sequence of events. That means each major threshold event makes several of the others likely by further increasing temperature—releasing methane from clathrates adds more heat, which speeds up the melting process on Greenland's ice, which elevates sea levels, and so on. Temperatures rise and reinforce themselves, leading to what some call a "hothouse Earth."[18]

In this case, human agriculture shrinks further and further as plants cannot produce because they evolved for a previous climate, or because their soils are either dried out or infiltrated by salt water. Humanity also loses the ocean as a food source as fish ecologies collapse. Food insecurity spreads broadly, followed by starvation. Potable water becomes more difficult to obtain. These changes to the natural world cause diseases to move and mutate rapidly, causing a range of illnesses and pandemics. The combination of higher heat and a worsened atmosphere add further stresses to this scenario's human body.

In response, human civilization might reproduce patterns echoing those of earlier crises. The provision of public and private services will become uneven at best, and corrupt or failed at worst. Some societies could harden into autocracies of permanent emergencies. Surveillance states backed by punitive force may cast themselves as logical solutions to chaos. Wars to protect or obtain resources are plausible, as is civil unrest. Huge numbers of people will migrate, seeking some basic safety; nomadic nations may appear out of that epochal movement. Given warfare, state oppression, migration, food insecurity, and diseases, the increasingly pressured human population will decrease decade by decade, emerging in the twenty-second century smaller than when it started the twenty-first. Beyond all of this physical misery, many people, nomadic or sedentary, will suffer extensive psychological trauma resembling that experienced by European and Asian populations during the Second World War.

As temperatures rise, habitable areas push ever closer to the poles, so we can envisage large human settlements in what are today Scandinavia, Siberia, and northern Canada and Patagonia, South Africa, and Australia. Ocean coastlines will be empty, but for scattered survivors, as we turn inland and to higher ground. New communities may burgeon on the slopes of the highest mountains, from the Himalayas to the Alps, the Andes, and the Pamirs. We have already referred to the difficulty in moving agricultural production to new areas, but it can be done to some extent. All infrastructure will need construction and hardening: roads, trains, airports, power lines, water supplies, sewage, communication from radio to broadband. Industry can also be relocated in these northern and southern extremes, yet we might not want to keep it going at its former level if it will keep the disaster going. In fact, if we cannot innovate our way to a new material basis for civilization, we could see a regression to ever earlier standards of living, with reductions to medical care, caloric intake, lifespan, and possessions. The mid-twenty-first century lifestyle could give way to that of the year 2000, then 1950, then 1900, and further backward still.

That assumes various groups have not made things worse. The historical record unfortunately provides us with many ways we could do so. Erecting higher boundaries between populations can gut trade and therefore economies. Wars fought to win advantage from one group for another can devolve into disaster for both. Reactionary movements can vitiate learning and the rights of many populations. Autocratic states can crush human expression, happiness, and well-being. Additionally, the realm of the hypothetical gives further mechanisms for worsening this worst-case scenario. As of this writing, the human race still maintains tens of thousands of nuclear weapons; it is not a stretch of imagination to envision ways nations wracked by collapse could find these of use or lose control of them to various interested and visionary actors. We also have practiced biological and chemical weapon creation for generations; this worst-case scenario could similarly see them in action. Some people alive at this time, those who have the luxury of studying ancient history, compare the present day to the collapse of Mediterranean civilization in 1177 BCE.

Combined, these threats present an existential threat to our civilization. In such a situation, it may be difficult to imagine how to preserve education and learning. Migrations away from coasts and formerly temperate areas leave behind most of academia's campuses, libraries, museums, and archives. To some extent, higher education will have to preserve itself in this great movement of peoples, reinventing the physical grounds for learning and research. Disaster-related research should be in high demand and also in the public eye, supported and monitored closely for whatever can help humanity survive. Teaching could follow suit, with topics deemed unrelated to the crisis receiving less institutional support and less enrollment. The undergraduate curriculum should shift to a disaster mode, emphasizing teaching practical skills for personal and community survival: medicine, agriculture, basic industry.

Speaking of institutional strategies might overstate what this dark future entails. Instead, we might see civilization collapsing so extensively that academic work is suspended.

At this point, we must think of the end of the human race. Which ends our very worst-case imagining.

Both of these very different scenarios and the middle one featured in previous chapters take us to the year 2100. That is the somewhat arbitrary endpoint for the present volume. The date is a popular one, both in climate change literature and the world at large, possibly because of the totemic power of a century's number. There is nothing otherwise special about the year 2100. Climate change will continue past that point, into the twenty-second century and, most likely, beyond as the amount of human-generated carbon dioxide persists in the atmosphere and the world continues pursuing the many associated effects. Academia, too, will—we hope—continue past 2100, if we manage to avoid this chapter's worst-case scenario.

I would like to close this chapter with another choice in addition to the one between best and worst cases. Looking past 2100 is more daunting than peering into the years before that point, as possible outcomes for so many variables pile up even further. Yet we can continue our forecasting by abstracting our focus to the highest human level, based on how we carry through the next several generations.

On the one hand, we have an alternative that starts with the Industrial Revolution and shoots past *Star Trek*. That is, we can draw on research that shows that the mass of humankind lived at a fairly consistent and stable level from the start of recorded history through about 1800. Considering a range of metrics—caloric intake, education level attained, average lifespan, the status of women, knowledge produced, a given society's ability to project power abroad—shows that preindustrial civilization was, roughly, remarkably stable and miserable over millennia. When Britain kickstarted the first Industrial Revolution and other nations followed suit, all of those metrics shot upward to a staggering degree. Powered by fossil fuels, the resulting civilization boosted humanity's status beyond anything experienced before. Yes, with costs including pollution and the climate crisis; yes, the benefits being unevenly distributed along a number of prejudices—but the historical rupture remains.

What if that trajectory persists? Imagine two more such centuries where such material progress and innovation continue at that amazing clip. Think of humans in 2300 living to two hundred years of age, when legions of diseases and another thousand natural shocks that flesh is heir to have been quashed. Technologically, if we began the year 1800 with horsepower, the first machine looms, and the earliest steam locomotive, then crested 2000 with the internet, antibiotics, nuclear power, the birth control pill, the first stage of artificial intelligence, and space travel, how much further will that development take us? Socially, if we consider the successive cultural changes since 1800—secularization, expanded civil rights for women, antiracism, socialism, ecology, nationalism—imagine how many more ways we can expand, enhance, and make more just the human condition. We can ask not only how this next human era will treat the earth, but what it will do with other worlds. Now we enter the realms of some science fiction, like the aforementioned *Star Trek*, and the visions of some futurists.

On the other hand, for the next centuries, we could instead commit ourselves to repairing the earth and restoring the balance between our species and its planetary home. We start with Bill McKibben's call in *Falter* (2020) to pause headlong progress for a time, recognizing our achievements but deciding to spend the future addressing their costs. Think of what results as civilization descends, gradually undoing the peak of industrialization. We remove carbon-generating machines and replace them with devices powered by wind, even muscle, offering a kind of refracted vision of the eighteenth-century world. Space travel is cut back to near-earth orbital satellites, those maintained in order to provide weather data; deep space exploration is too costly an investment for too little a return, too dehumanizing and distracting from the Earth's realities, as Naomi Klein argued. We devote resources to restoring and expanding the natural world, growing rewilding sectors at the expense of human settlement. Humans pile up more closely together in dense cities, our total population shrinking after a mid-twenty-first century peak. Following Donna Haraway's vision, we produce fewer children while instead partnering more with animals, both individuals and species, aiming to build kinship relations with other beings rather than reproducing our DNA. Humanity repairs mountains, rivers, and oceans. Our work is restorative, rehabilitative. We consume less each generation, constantly seeking to live more lightly on the earth. Instead of expanding the human enterprise into the university, we limit it to minimize planetary harm; instead of amending the earth, we mend it.[19]

The gap between these two visions is vast. The role of higher education in each is staggeringly different. The choice between them is one we may make by how we conduct ourselves for the next two to three generations. That choice may be unconscious and emergent or deliberate, openly and strategically made. Academia can play a role in that species-wide decision. What choices we can make toward that end are the subject of the next and final chapter.

CHAPTER 7

What Is to Be Done

Perhaps the great awakening on warming has already happened—or keeps happening and keeps being forgotten, among other reasons so that we can continue to believe we stand just at the threshold of climate suffering rather than well beyond it. But the great awakening on adaptation probably still lies ahead of us. Or maybe that "permanent emergency" is beginning right now. —DAVID WALLACE-WELLS

Achieving global net zero CO_2 emissions is a requirement for stabilizing CO_2-induced global surface temperature increase, with anthropogenic CO_2 emissions balanced by anthropogenic removals of CO_2. —IPCC 2021 REPORT

Time is no longer on our side. Let's use what time we have more wisely.
—MICHAEL MANN AND TOM TOLES

At this point in the book, the reader might experience a powerful sense of gloom, even despair. The preceding chapters present a picture of a vast, persistent, and brutally destructive threat already starting to hit academia around the world. The two great tasks the climate crisis presents humanity—to reduce greenhouse gas emissions to zero and to adapt civilization to a new world—are enormous, historically unprecedented, astonishing in scope and complexity. Confronted with them, some may experience decision paralysis. The idea of attempting to preserve academia for another century could overwhelm us. Others may cede the struggle completely, as does the Dark Mountain group, or to view climate change, as writer Jonathan Franzen recently put it, as "a lost cause."[1]

Academia is also not so fortunate as to confront the climate crisis as its sole challenge. Higher education worldwide confronts a swarm of additional pressures. The world we inhabit presents us with the demographic transition, changes to the labor market for graduates, inequalities and violence by multiple categories (class, race, gender, sexuality, religion, geography, ethnicity), changes to public attitudes toward higher education, political directives, warfare, and, of course, disease. Academic institutions struggle internally with their own economic sustainability, curricular and pedagogical development, supporting research, integrating technology, hiring and supporting faculty and staff, supporting students, and many more issues besides. To an extent, each of these challenges appears independent of the climate crisis and must absorb some of our resources. At the same time, each can connect with global warming and make the picture more complex still. Each institution's experience will vary, of course, but colleges and universities worldwide will have to grapple with the climate emergency, perhaps in escalating intensity over the next eight decades—in addition to all of those other problems. Academics as individuals and as participants within institutions will have to determine priorities and apply resources accordingly, which is something we have always done. Yet the Anthropocene appears in addition to our past efforts, and in some crucial ways at a far greater scale.

In the face of all of this, colleges and universities can appear to be small and fragile entities, massively out of their depth in the crisis. To consider academia as a human institution from which humanity in general might draw strength may seem starkly unjust, especially for an academic sector often reeling from financial, political, and epidemiological pressures. Viewing higher education as a space from which to mount a positive response to the unfolding catastrophe may seem woefully inappropriate. Individual academics can feel especially minute, dwarfed by the challenge, overwhelmed by its enormity. These are deeply human and valid responses, and we will require time to work through them. Yet we must also react practically and strategically if academic institutions are to survive, much less thrive, for the rest of our century. Further, colleges and universities *can* play a decisive role in humanity's response. As we've seen in this book, higher education can contribute so much, from our research mission to our teaching, from our local community service to our engagement with the world.

Where to begin?

We can start with the question of motivation. Which academics would be interested in climate action? It should be clear by now that some number of people within and adjacent to higher education—faculty members, staff, students, government officials, nonprofit officers, business employees, journalists—will view the climate crisis as a gigantic or existential threat and want to take steps accordingly. A good number already do so. It seems likely that their numbers will grow in the years and decades to come, as the crisis deepens and expands. There will also be people who see the university's Anthropocenic response as an opportunity for them to pursue unrelated goals: to boost their own reputation, settle personal scores, build empires, make more money, enable a career shift, and so on. It would be naïve to imagine otherwise. As academia's climate response grows, both of these populations (and there is overlap, of course) may well expand. Climate crisis activism can become its own movement, lobby, or constituency on and between campuses.

We should also add to our answer a kind of personal-institutional attitude. The climate emergency presents to colleges and universities a way of rethinking, reimagining, and redesigning the entire academic enterprise. It opens up a possibility space perhaps larger and more ambitious than we have seen in some time, since so much of our institutions are in play, as I hope this book has demonstrated. There is a defamiliarization at work, in literary critical terms, as global warming causes us to rethink and resee much of the academy, often in new ways and in different combinations. The same is true for the academy's role in the world, both locally and globally, as we saw in chapters 4 and 5.

I do not want the reader to think that I am cheering on the Anthropocene as a delightful opportunity for individual and institutional growth, although the latter will occur. This is a dire threat with terrible human costs in the wings; readers unconvinced of this should reread chapter 6. Institutions and individuals may seize the opportunity to do foolish or cruel things, as we learned from recent crises, such as the COVID pandemic or 2008's financial crash, and as even a casual glance at history will demonstrate. Academia's response to the crisis over the next eighty years will likely include many missteps and heinous acts. It would be naïve to imagine otherwise.

Yet for all of that, the climate threat remains, and it will worsen. Academia should not look away from it. We should reimagine and redesign our institutions, not only in response, but proactively. We can play a larger role in the crisis in our communities and in the world. We have a vast capacity for thoughtful, creative, and humane institutional and personal imagination. As Georgetown University's Red House initiative declares, "A university can reinvent itself."[2]

Let me now bring into our discussion two very different views. They aren't mine, but ones I can ventriloquize based on conversations I've participated in over the past decade with academics around the world. The first voice is that of a professor or senior administrator who has considered some of the evidence, then concluded: "I can see how global warming will endanger certain universities—but not ours. We are protected by our geographical location, far from deserts and oceans. Our government/donors/church will assist us with what we need, should anything bad occur over the next eighty years. The broader civilizational issues you describe are, of course, important, but we are only one academic institution, and we're also addressing other major issues at the same time. The faculty we have can research and teach enough climate change to satisfy students and accreditors. There's really nothing else we need to do."

The second voice I'd like to represent belongs to another senior campus leader, who asks: "What if we're already doing enough? We have faculty researching climate change and have done so for years, so we are strong on that front. We also teach the subject, both in the form of majors and graduate programs as well as a campuswide commitment to sustainability. Our campus architecture? Well, some of it is dated, but the key buildings are LEED certified and we have capital plans for improvement. We purchased insurance policies that are strong enough to take care of challenges. The experience of surviving COVID-19 taught us well to be resilient in an emergency. What else must we do?"

Those are good questions from a thoughtful and active president, professor, chancellor, board chair, or other senior leader. They describe progress that we should recognize and appreciate. Yet there is still more to be done, depending on an individual institution's situation and how the next generations of climate crisis pan out. Recall the three-part model of threats introduced in the introduction to this book. Primary stresses can strike campuses away from seas and deserts, and that have purchased insurance—strike with some unpredictability, and perhaps more often than in the past. Secondary stresses can be hard to anticipate, as they work through complex systems: ecologies, economics, geopolitics. And the tertiary stresses can approach in any number of forms, based on the sheer variety of human sociopolitical imagination. Consult chapter 5 for a sketch of the ways the world might impinge on a college or university. In other words, a given academic entity can make progress and avoid some leading dangers, while still being subjected to dangers.

Even more remains to be done if an academic institution wants to not only survive reactively but proactively participate in the global movement to mitigate and adapt to the heated world. Colleges, universities, systems, alliances, and other organizations can play a leading role in this ultimate crisis. We can serve as exemplars and inspiration, demonstrating practical ways to mitigate and adapt, even at the small scale of a single school. Higher education can advocate for global transformation through many different mechanisms, from faculty and staff acting as public intellectuals to lobbying businesses and governments with which we work. We can contribute capacity to mitigation and adaptation efforts, such as assisting civil engineering projects with our knowledge and labor. As we reinvent higher education, we can help civilization reinvent itself.

Given these questions, answers, and possibilities, let us outline steps to consider, starting with the short term.

What to Do in the Short Term?

Colleges and universities should consider their organizational stance toward the unfolding Anthropocene. What attitude would best mobilize their resources for survival and development? I have previously argued that a futures orientation best positions an institution for uncertain years and likely challenges in general and think that advice holds in this context. This should include an attitude favoring experiments and explorations, which is not necessarily as easy to adopt when a campus sees itself as under threat or being charged with guaranteeing a heritage. A posture dedicated to resilience should be essential, even if the term is overused.

Internally, campuses may consider forming or dedicating a unit to the climate crisis. The recent COVID experience can be useful here, in that institutions had to assemble groups who gathered intelligence, used it to shape policies often on the fly, communicated policies with clarity and transparency, and were ready to shift operations at a moment's notice. Such groups may begin on an informal basis but may develop into permanent bodies. They should inform strategic planning over time. Governance implications may also follow the coronavirus pattern, depending on the institution. That is, some campuses formed an emergency management leadership group, while others partnered closely (voluntarily or otherwise) with local or national public health authorities. In some situations, senior administration expanded their internal governance role at the cost of faculty's; this dynamic will doubtless occur again and form the ground of political contestation.

We can find early examples of this kind of organization already in use. Eckerd College's president organized a resilience plan in combination with senior staff, some faculty, and external consultants. That became a group, a "task force of executive college staff and faculty . . . includ[ing] President Eastman, Sparkman and several other administrative staff and faculty members." The University of California, San Diego set up a Climate Action Planning Group with many campus offices and stakeholders to create a plan. Members represented facilities, engineering, sustainability, and other units. The plan the group developed addressed current campus GHG emissions, future scenarios, and mitigation strategies. However, the UCSD plan also had minimal faculty involvement.[3]

Such a campus group may also rely on external support. This can entail working informally with peers and networks, as well as utilizing formal relationships with associations and governments. External design groups may prove useful, as Eckerd College found. Such work can involve a participatory, iterative process with many institutional populations.[4]

External relations include government relations, of course. On the local level, individual campuses may conduct more work with their immediate communities. As we saw in chapter 4, this could take the form of increasing collaboration between members of the academic institution and local groups and institutions. It can also appear as friction or hostility. Negative and positive town-gown interactions can occur simultaneously or in close sequence, especially over years and decades. Doing so well may entail assigning community relations functions to administrators and/or faculty, or the creation of detailed positions or offices. At a broader scale, we might anticipate government relations officers devoting more time to climate change issues. This could involve taking up GHG reporting roles or establishing positions focused on that function, as we saw in chapter 5. Of course, governments of all levels may proactively work with colleges and universities for their own purposes. As we also saw in chapter 5, we can envision regulations and policies that include or specifically target academic institutions. A government might require postsecondary institutions to teach victims of climate disasters or migrants. It could nationalize some research it deems relevant or prohibit certain teaching. A campus leadership group will have to be very attuned to government attitudes and actions, seeking to protect its institution while maximizing benefits and opportunities from the relationship.

Colleges and universities should also consider meaningful work on the representational or symbolic layer. I do not mean this in a cynical or shallow way, but instead to draw attention to linking actual climate work with how the world perceives an institution. It is quite possible for a postsecondary institution to conduct research on climate change, transform its pedagogy, and mount all kinds of public work without making a dent in its reputation. Further, since so much social interaction and political transformation takes place at the symbolic level, for better or for worse, interacting at that level can be productive for helping us grapple with the climate disaster. To this end, colleges and universities may change their physical appearance to look greener or more climate committed. Huge wind turbines, roads replaced with bike paths or lined with solar cells, new buildings that are net negative emissions and look very different from the rest, and so on can make a clear statement of an institution's priorities. A more literal statement can come in the form of an institution's mission statement. These can easily be general boilerplate with little practical application, yet perhaps language about grappling with the Anthropocene will signal that a campus has some resourced intention along that line.[5] Universities or their executives could also issue public statements to influence public opinion and decision-making about climate issues, drawing on their institutional reputation and analytical powers. They can come together to encourage each other in taking concrete steps toward decarbonization. For example, the Race to Zero declaration bound signatories to these actions:

- Pledge: Pledge at the head-of-organization level to reach (net) zero GHGs as soon as possible, and by mid-century at the latest, in line with global efforts to limit warming to 1.5C. Set an interim target to achieve in the next decade, which reflects maximum effort toward or beyond a fair share of the 50% global reduction in CO_2 by 2030 identified in the IPCC Special Report on Global Warming of 1.5C.
- Plan: Within 12 months of joining, explain what actions will be taken toward achieving both interim and longer-term pledges, especially in the short- to medium-term.
- Proceed: Take immediate action toward achieving (net) zero, consistent with delivering interim targets specified.
- Publish: Commit to report publicly both progress against interim and long-term targets, as well as the actions being taken, at least annually.[6]

It's important to remember that much depends on the specific nature of a given postsecondary institution. Its geographical location radically situates it within the climate crisis itself. Its institutional nature—research university, technical institute, military academy, religious college—fundamentally shapes an institution's choices and parameters. Public colleges and universities are intertwined with governmental policy to varying degrees, while private campuses are more closely tied to private funding and governance sources.

We now turn from questions of institutional leadership to matters involving the physical campus, and in this area, many options and challenges loom ahead. We can start with a university's computing environment. How can that infrastructure survive climate change–driven dangers, both on- and off-site? One option is to build redundancy by hosting and backing up data and applications to off-site (i.e., cloud) hosting. Another is to participate in intercampus collaborative hosting projects like LOCKSS (Lots of Copies Keep Stuff Safe) or CLOCKSS (Controlled LOCKSS), where members agree to host other campuses' data, creating a resilience backup network in case of disasters.[7]

At the same time, campuses will most likely rethink their IT strategy and operations in terms of their carbon footprint. For example, does an institution reduce or directly block the use of some technologies? It would be unsurprising to see a university ban bitcoin mining, given the enormous (and growing) carbon emissions involved in that process. Perhaps campus policies will restrict some AI research and usage, such as training software on huge datasets, because the cumulative computation involved means generating a large amount of carbon emissions throughout the process. Alternatively, an institution might want to restrict such computational intensity because campus leaders see it drawing green energy that others could use to purposes the institution would prefer. Hardware provisioning may shift to easily upgradeable devices, rather than replacing them entirely. Campuses may encourage—or mandate—green computing practices in general. Governments, local or national, could well encourage or require the same, and nonstate partners (companies, nonprofits, civil society organizations) follow suit. To determine such policies, academic institutions must measure computing's impact, which requires research and the creation of metrics with new techniques. Conversely, some institutions should prepare to defend the carbon costs of conducting computationally intensive research; in certain contexts, climate research may be the easiest to protect. Using a mixture of distributed sensors with analytical software to understand campus building performance may be another justifiable use of computer resources, even to the point of creating a "digital twin" that simulates the academic space. Beyond immediate computational use of energy, how will a college or university account for the social costs attending on environmental stresses?[8]

Computing is just one part of a physical campus's energy needs. There are many other steps an institution can take to reduce its carbon footprint. On a small scale, switching lights to LED is an established practice, already seen in related cultural institutions, like opera houses. Replacing appliances that use chlorofluorocarbons and hydrochlorofluorocarbons, notably refrigerators and air conditioners, is quite feasible. Replacing static windows with electrochromic smart glass, which changes the amount of heat and light it allows through based on computer instructions, is an emerging technology worth exploring. Improving overall energy efficiency through insulation, replacing some materials, and more closely aligning energy consumption with actual use is a relatively easy win.[9]

At a broader level, we may rethink, redesign, and even overhaul the built environment of a physical campus, as we saw in chapter 1. Some buildings may be fortified against storms, fire, and particles. Others could be renovated to reduce greenhouse gas emissions. Still others could be relocated, either to a more favorable site on campus or to another spot in the community. Consider the case of Washington College, as described in Lee Gardner's groundbreaking 2019 reporting:

> The student union, which was inundated in 2008, now sits behind a 360-degree flood wall, more than 10 feet high on the river side. The university also invested in nine miles' worth of large-scale interlocking prefabricated barriers that can be quickly moved into place and filled with sand by bulk loaders to protect other sites.

Some sites can be protected only so much. The building that houses the theater-arts department at Iowa sits right on the river bank, and the 2008 flood deposited about six inches of river-bottom sludge on every horizontal surface on the first floor. Iowa has since prepared the building for deluges to come. All of the mechanical systems were moved up from ground level. The university redid the ground floor with durable triple-layer plaster, raised electrical outlets, and minimized the activities held on that floor.

Campuses in vulnerable areas also need to design against fire. Those in arid areas as well as regions where droughts are increasing, such as the Universidad de San Carlos de Guatemala, should plan for the likelihood of advancing flames. Some have already done so and can be learned from. The University of California, Los Angeles, is changing the mix of trees on its grounds for this reason:

> The drier weather has begun killing off some species of trees planted around the campus, such as non-native Canary Island pines. As part of a master landscape plan, the university is now looking at replacing some trees with better-adapted native species—and getting rid of its eucalyptus trees altogether. The aromatic oil that suffuses them means "eucalyptuses are quite flammable," says Nurit Katz, chief sustainability officer. "The large number of eucalyptus [*sic*] is part of why the Oakland Fire in the '90s was so severe."

Near to UCLA, Pepperdine University has a very extensive fire prevention plan:

> Only about 300 acres of the 800-acre site have been developed—the rest stand as a buffer. Closer to the developed core, the university maintains a 200-foot zone cleared of flammable brush around all structures. Phil Phillips, vice president for administration, has watched fire race toward the campus five times, but "when it hit that brush-clearance line, it would just kind of fade out, peter out, over and over and over."
>
> The Mediterranean-style buildings themselves are faced with stucco and roofed with terra-cotta tiles, both fire-resistant surfaces. They have no eaves, which are among the easiest entry points for a floating ember or stray flame. The buildings are arranged in clusters, but even as the campus has

expanded, they have been spaced widely enough that fire couldn't easily leap from one to the next, and firefighting equipment could be maneuvered between them.

Pereira even built in a campus water-reclamation system that the university uses to feed its two-piece fleet of firefighting equipment—and its public-safety officers are cross-trained in wildland firefighting.[10]

Renovating old buildings and rethinking new ones will bring to mind the options discussed in chapter 1. Redesigning structures to be carbon neutral is one large step, involving careful building siting, improved insulation, changes to windows, overhauled appliances, and new materials. On the construction side, we may make more use of recycled materials, as well as tending to refurbish present buildings instead of replacing them totally.

There are many opportunities for innovation on this score. For institutions located on bodies of water, elevating entire grounds may be the best option. Bringing together sufficient dirt or landfill to the sea might reduce mountains or create vast new pits inland. Building massive seawalls is another possibility, one already established around the world, from Holland to Guyana and Puducherry (formerly Pondicherry). These, of course, run the obvious risk of being leaky, overtopped, or broken. They run the less obvious but no less threatening risk of trapping flood waters on campus, if the walls survive a flood, as we have seen with Hurricane Katrina in New Orleans. Another option might be floating buildings, which have already been deployed in São Vicente and parts of Lagos.[11] Campuses may loft buildings and walkways on pillars or stilts in order to let waters pass harmlessly beneath them. In any event, what is to be done with landfills an institution uses, either those located on campus or elsewhere? To what extent is a college or university willing to expand its sense of recycling and waste cleanup?

There are still more options when it comes to building design. To pick one source of inspiration, the Living Building standard, currently in version 4.0, offers a very challenging yet inspiring set of requirements. They include growing food on-site, protecting land for animals on- or off-site, offering support for bicycle and electric vehicle transportation, embracing biophilic design, and enabling access to outside air and light. Materials must be locally sourced to certain degrees and avoid containing any "Red List" chemicals. A building created and maintained to this standard must also "supply one hundred percent of the project's water needs through captured precipitation or other natural closed-loop water systems, and/or through recycling used project water, and all water must be purified as needed without the use of chemicals."[12]

The COVID-19 experience may yield further ideas for academic building design in the Anthropocene. The importance of safe and healthy ventilation is encouraging schools to improve filtration and circulation systems. Larger and a greater number of doors and windows may fulfill that purpose as well. We may see a move toward more spaces with direct outside exposure. The pandemic may literally change the face of higher education, a change climate planners may continue. At the same time, we must remember that every institution occupies a different ecological niche with its own constellation of threats. Different types of institutions have varying capacities and responsibilities. A religious university, for example, may be bound to its faith's climate strategy, while a public campus has to navigate state politics. (See, for example, a call for Catholic institutions to use their land for power from Dan DiLeo, faculty member and director of Creighton University's Justice and Peace Studies Program.) It may be that small pilot projects are the safest way for an institution to advance. In 2005, a group of Leuphana University Lüneburg students and staff set up a pilot solar project. In so doing, they navigated university power structures, financing, and maintenance: "The initiative's members . . . had to acquire knowledge and skills to build up and run a private initiative. They also had to obtain and coordinate internal and external knowledge on photovoltaic technologies." By 2009, Lüneburg embraced the installation for future maintenance, and it sounds like a successful pilot that suited its situation well. Ultimately, each campus has its own distinct nature and has to assess its individual risk exposure:

> The myriad effects of climate change can be very localized. A university in Manhattan, for example, might have to deal with rising sea levels. A college in

> Denver may be faced with colder winters. A warmer planet generally means more flooding, wildfires and extreme heat. Colleges at this stage often also assess the politics and infrastructure of their home location. Is public transit robust? What are the bonds like between the college and the county? In some cases, administrations may choose not to release assessments that expose the vulnerabilities of their institution, [Alex Maxwell, senior manager of climate programs at Second Nature] said.[13]

Higher education institutions should also reconsider their inherited sources of electrical power. Campuses may increasingly install solar panels for their own use and in different forms. Alberta's Red Deer Polytechnic now generates 16 percent of its electrical needs through thousands of solar panels distributed across its campus. That includes a new residence hall that integrates hundreds of panels into its structure, not just on top of its roof, but on walls and other surfaces, "form[ing] part of the building or the building envelope." In 2021, Grinnell College broke ground on a standing solar farm installation designed to provide one-third of campus power needs. Some academic solar projects may involve collaboration with the local community, as with Vermont's White River Community Solar project driven by a graduate school (Vermont Law School) and a nonprofit (Building a Local Economy). In addition to different economic and format structures, different types of solar cells are becoming available. Besides different ways of siting panels, there are also varying materials, such as crystalline silicon and cadmium telluride. Beyond solar, larger institutions may consider hosting wind turbines, depending on local meteorological and political conditions, while smaller ones may site microturbines.[14]

Other power sources may become available over the rest of this century, depending on technological development, finances, and an individual institution's situation. Geothermal power is, for now, more affordable in certain areas that have underground hot water close to the surface (for example, in Iceland). We may have campuses decide to take advantage of this power generation method as carbon costs rise and as—and if—geothermal becomes more flexible and affordable. For colleges and universities near substantial bodies of water, hydropower may become attractive as current and tidal technologies improve. "Low-head" hydropower could work for those campuses located next to fast-moving rivers. One instance of this approach appeared in 2021, when Berea College successfully installed a hydroelectric plant on the Kentucky River, which should provide 50 percent of that institution's power needs. Perhaps some institutions will deploy or outsource anaerobic digesters to turn food and agricultural waste into natural gas as a transition fuel, as Middlebury College has done. Concentrated solar power, whereby many mirrors focus sunlight on a fluid container to heat the contents to turbine-driving steam, may become more efficient and therefore attractive, especially in regions with the highest amounts of sunlight. Biomass can provide heat as well as electricity, as a University of British Columbia project has demonstrated. Nuclear power may appeal, either in the form of very small reactors on site or, more likely, as an off-campus outsourcing strategy, especially as the climate crisis deepens. Naturally, each of these power sources has its own issues, from competing with other needs for certain sites to political opposition, noise pollution, and significant maintenance or decommissioning costs. Each may appear on campus (or be outsourced), serve for years to generate power, then give way to successor power generation systems. Each can figure into campus emission audits as well.[15]

When Universities Draw Down Carbon from the Skies

All of these physical campus measures involve academic institutions fulfilling one of humanity's great Anthropocenic tasks, cutting down our GHG emissions. Campuses may also decide to participate in a second such task, drawing down previously emitted GHG from the atmosphere. The most popular forms of this are afforestation and reforestation. La Universidad Autónoma de Madrid planted trees in a carefully selected local area, collaborating with another university and the city government in choosing a tree species, developing a road, determining tree density, and paying bills. These plantings can be done badly (trees planted in bad soil, too closely together, or in the wrong climate for a given species) and their benefits overestimated (it takes years, often decades from planting to realizing results), but done properly, this is a strategy with positive long-term results. There are as many ways to expand tree count as there are microbiomes and forms of human ingenuity. Mangroves might be a good choice for campus grounds experiencing saltwater intrusions. A campus might select or genetically modify trees to have lighter colors in order to not only gather carbon dioxide but also to reflect back more sunlight. Adding trees and other plants to campus grounds often suits popular and historical impressions of academic styles, such as the cliche of "ivy-mantled tow'r[s]" in Thomas Gray's famous line. More foliage has the additional benefit of improving mental health, according to biophilic design thinking.[16]

A second strategy involves installing DAC devices on campus. We could imagine seeing such machinery in small numbers and at likely locations to start with, such as near engineering or environmental studies offices. DAC numbers and scale may well increase as the century progresses, growing in quantity to banks and installations, each device improving in efficiency as the technology improves. Perhaps climate-minded colleges and universities will compete with one another on their carbon sequestration capacity; perhaps institutional ranking schemes will account for this metric. Anand Kulkarni of Victoria University has suggested the creation of a resilience ranking to be hosted either by a published, third-party entity or by campuses themselves. How to set this up can involve many specifics—Kulkarni recommends considering faculty research output, student value, risk management, consistently following published values, diverse revenue sources, and demonstrated flexibility—and it is easy to find many critiques of ranking systems. The key point is opening up climate operations as a domain for institutional competition, with DAC as a measurable, contestable metric.[17]

A third carbon sequestration approach involves agriculture and campus grounds. "Enhanced weathering" works by amending soil with materials like basalt, wollastonite, or olivine, which, ground up to a very fine powder, can be distributed over certain areas to reproduce the carbon-gathering capacity we see in rock weathering. Other materials could be used for this purpose. Academic researchers from Sheffield to the University of Guelph have been experimenting with enhanced weathering: "While . . . researchers apply basalt to hemp in New York and to alfalfa and olive trees in California, scientists working with the University of Sheffield's Leverhulme Centre for Climate Change Mitigation in the U.K. are spreading basalt on cornfields in Illinois and on sugarcane in Australia. In Ontario, Canada, researchers are applying wollastonite from a nearby mine on soybean and alfalfa fields." As Elizabeth Kolbert recommends, "Basalt could be mined, crushed, and then spread over croplands in hot, humid parts of the world. The crushed stone would react with carbon dioxide, drawing it out of the air."[18]

As with any technology, these three carbon-capture techniques bring with them ramifications, side effects, and knock-on consequences. Banks of trees or DACs will change a campus's physical profile. Each may require significant amounts of water to function, which could necessitate expanding institutional water supply, which may in turn stress local hydrological systems. Enhanced weathering may require setting up ground rock stockpiles or conducting the grinding on site. Campuses might build or expand greenhouses using drawn-down carbon dioxide to aid plants in growing food. Storing sequestered carbon might take place on campus, underground, or at an off-site location, each of which will require monitoring.[19]

Besides carbon dioxide, campuses are also likely to pay increasing attention to their role in controlling methane, currently the second most damaging greenhouse gas. This can involve rethinking landfill use, such as by more extensive recycling, repurposing, and reuse. It may involve adding new sewage treatment methods. As with carbon dioxide, this means rethinking human diets as they appear on campus—that is, food service plus public health communication. Taking steps to control food waste can slow methane creation. A methane strategy should involve changes to the way any animals are raised on campus grounds, if there are any. A methane plan also overlaps with a carbon reduction effort, since fossil fuel production also emits methane—35 percent of the human total, according to a United Nations report.[20]

Many of these plans and alterations are costly, of course, and campus finances may have to change in other ways. For some institutional leadership, redoing their physical structure appears to be too expensive, which leads them to purchase carbon offsets instead. This may be a short-term solution, one that divides faculty, students, and staff into supporters and opponents. That latter population may also press on institutional stewards to divest endowment holdings in GHG-intensive businesses. Student activism may play a key role here, drawing on previous divestment campaigns, as Naomi Klein has argued. Institutions could take a similar stance with regards to charitable gifts from certain entities, in the form of refusing carbon-burning money. For example, Britain's Russell Group universities received "more than £60m in research and teaching funding from companies in the coal, oil and gas sector" during the 2015–2020 period. Will activism—either within academic institutions or beyond—call for that practice to end? How will those charged with fiduciary responsibility respond, especially given rising demands on their budgets? Several American universities, including Harvard and Boston, recently divested from fossil fuel industries. Additionally, colleges and universities will likely revise their insurance policies. As climate pressures rise—again, unevenly, depending on a given institution—negotiations over pricing and coverage will likely heighten.[21]

Overall, we might expect to see colleges and universities set decarbonization targets using various metrics and timelines. At this point, there are many options, including the question of whether to aim for net zero or negative carbon emissions. Academic leadership can also alter targets, much as their governmental, business, and nonprofit peers have done. For example, Southern New Hampshire University recently advanced its target date from 2030 to 2025. Those peer organizations may also seek to influence or compel campus decarbonization and methane reduction plans. Scoping out just what constitutes a carbon or GHG footprint requires a major decision at three and more levels and striving to be carbon negative across all of them. The nonprofit Carbon Disclosure Project (CDP) offers three options for defining an organization's footprint, in Solitaire Townsend's summary:

> Scope 1—Direct emissions (from a company's own vehicles and facilities)
> Scope 2—Indirect emissions (from the electricity that a company purchases)
> Scope 3—All other indirect emissions (from supply and even in consumer use)

We can see a college or university choosing one of these and defending that choice. Townsend actually thinks these are too limiting, and calls for a new level, "Scope X," or "work that restores and regenerates, that rebuilds the foundations of healthy ecosystems and thriving communities, that takes responsibility for system level emissions . . . A Scope X company would act on carbon far beyond their direct or even indirect impacts, and instead accept the challenge of system change." Townsend is referring to corporations, especially the largest ones, but we can imagine some academic populations seeing Scope X as fitting their sense of mission in the community and world.[22]

The Question of Academic Travel

A key source of those Scope 1–2 and also 3 emissions in that academic setting is transportation, of course. As noted in chapter 1, that is one area colleges and universities seem likely to rethink and redesign. We have already seen some institutions replacing the fossil fuel–burning campus shuttles, carts, cars, and other vehicles that they own and operate with electrically powered ones. La Universidad Autónoma de Madrid took steps to expand its accessibility to bicycling, adding secure storage centers, making bike rentals available, and adding bike paths and lanes totaling "almost 100 km." Australian National University created the "Timely Treadlies departmental bike fleet," which has "reduce[d] motor vehicle use." UCLA recently vowed "to run a completely carbon neutral fleet by 2025." This parallels similar moves in primary and secondary schools.[23]

Beyond the makeup of vehicles campuses own is the more complex question of academic travel in general. We may face this issue for a generation, as at least one major airplane manufacturer states it will be using conventional fuels, which emit a great deal of carbon dioxide, through 2050. Student travel raises its own set of questions. First, will we come to view remote learning at an institution located in another country to be on par with physically traveling there? Or will students turn to low- or no-carbon transportation for international study, such as trains, ships, or mechanisms to come? Second, how will an aggressive decarbonization agenda impact study abroad? One can imagine a college buying carbon offsets for the next few years, arguing that the personal and social benefits of that course of learning justify the action. Or will students and/or program leaders prefer study that requires shorter travel distances, or perhaps aim for a combination of fewer trips with deeper cultural dives, as Hans de Wit and Philip G. Altbach have called for? For example, their take on Europe's Erasmus+ program:

> More than 10% of European students participate in the flagship Erasmus+ programme, moving around the continent and beyond for periods between two months and one academic year. This is the most extensive mobility scheme in the world. The European Union, which advocates a more aggressive policy on climate change, should also take measures to reduce the environmental impact of this programme, only allowing more climate-neutral forms of mobility and exchange of students, teachers and administrators.

Alternatively, will study abroad anchor itself on what Jenny J. Lee and Ola A. Lundemo refer to as "regenerative partnerships"? These voyages would focus on improving a community's sustainability and resilience in the face of the climate crisis and doing so with preexisting study abroad benefits: learning another language, exploring a culture, and practicing mobility, independence, and interdependence. This might have curricular impacts at sending institutions: "A regenerative approach to international education also involves eco-consciousness raising, such as incorporating environmental education and research into other disciplines. Infusing environmental topics into and across the curricula, sponsoring interdisciplinary meetings to address climate change and supporting nature conservation research are obvious places to start."

Third, will campuses with significant residential student populations revise their academic calendars to reduce the frequency of holidays and vacations and thereby cut down student travel? It is not difficult to imagine new academic calendars in the climate emergency. We potentially saw academic institutions' ability to make such changes when colleges and universities altered their calendars in reaction to COVID-19's successive waves.[24]

Faculty and staff travel may fall under similar scrutiny. Given the environmental costs of professors and staff flying, especially longer distances, we might expect charges of flight-shaming and calls to simply cut back such trips. Those calls may come from within the academic world as well as from the outside. They can easily be politicized, as when (for example) right-wing politicians criticize environmentally minded politicians for extensive jet travel. Rather than reducing such professional activity outright, we can attempt to replace some or all of it with digital means. The COVID-19 pandemic taught us many ways this can be done, both badly and well. Yet flipping a conference, meeting, or on-site research experience online may not be sufficient. We can also seek to reduce the GHG footprint of virtual experiences. Grant Faber published a framework within which virtual event participants can cut back their emissions, from use of electric lighting to server traffic. Reducing emissions may also entail new arrangements with on-site hosts, including hostels, local transportation, food service, and other parts of local society.[25]

Academic travel may learn from some of the thinking around green religious pilgrimages, which seek to preserve the spiritual experience while reducing the overall GHG footprint. James Mills offers suggestions that might inspire non-pilgrimage-based academic applications: "Pilgrimages could be local and based on walking, therefore limiting the use of fossil fuels for mobility. Pathways could simultaneously serve as wildlife corridors, connecting fragmented habitats and making movement safer for birds, mammals and amphibians. Green cemeteries could be integrated into the pathways further increasing natural habitats while serving a human purpose."

Mills goes on to find an additional benefit to these new practices, one that resonates with some climate pedagogies discussed in chapter 3: "Intimate contact with local landscapes while on a pilgrimage can help develop a stronger sense of place, which often leads to greater concern and stewardship of the environment." Perhaps a reduction in physical travel will enhance academics' sense of their campus and its immediate community. Perhaps that local community might seek to encourage this new mode of academic work: Juliet Osborne, chair in applied ecology and leader of the environment and climate emergency working group at the University of Exeter, said that "academics could be better at sharing the outcomes of conferences to reduce the number of scholars attending each event and universities could highlight the health and equality benefits to reducing travel."[26]

We must also keep in mind that travel technology will surely change as the twenty-first century progresses. Recall that the first heavier-than-air flight was just over one century ago (1903) and consider just how much innovation, expansion, and development occurred in that span. Looking ahead, academics might anticipate revised designs for jet aircraft. Less GHG-intensive fuels may appear, including some that include biofuels or synthetic aviation fuels (SAFs). More efficiently burning fuels would help, as would improvements to aerodynamic design. Planes may eventually fly powered partially (as hybrids) or entirely by hydrogen or batteries, the latter considered to be functioning electrical aircraft, at least for short distances, where smaller craft and slower speeds are accepted, since batteries are heavy (for now). We may also see airships and blimps return to play a key role in travel. Their motors run by electricity, rather than by burning fuel, and can be powered by solar or batteries.[27]

The New Research Mission of Universities

Some of that research and development will come from the academy, and here we return to the research enterprise, how it might be transformed by the next several generations, and how we can best anticipate and plan now. To begin with, it seems likely that many universities will increase their support for research into climate change with professional development, hiring faculty and staff, and creating new academic units. As we saw in chapter 2, many academic fields can research aspects of the climate crisis. Such scholarship can contribute to our common understanding of the crisis, which is a clear case of academia serving the public good. This research can play out across the full breadth of the Anthropocene as a topic, from analyzing changes to ocean currents to modeling how societies may respond to increased temperature. In order to support this research agenda, some institutions will be faced with the additional cost of some physical research that runs into climate obstacles, like gathering ice cores from decaying glacier fields or conducting fieldwork amid advancing prairies. This could involve temporarily hardening certain sites under extreme conditions or building very high-fidelity digital twins of the source material.[28]

Academics may also carry out not only research but also combined research and development of new technologies, services, and more. Purdue University faculty have developed a complex new paint that helps materials cool off more rapidly, based on barium sulphate, nanoparticle films, and nanocomposite paint. Other researchers have developed new structures and designs for floating buildings. Still others continue developing alternatives to meat, such as vat-grown food built up from the cellular level. Some explore alternatives to oil-based plastics, like mycelium. Oceanic power, both tidal and current based, calls out for rapid R&D, as it has fallen behind solar and wind in terms of efficiency and price. Faculty, staff, and students might work on improving the efficiency and price of devices that obtain water from the air, which could be vital hardware for drought-afflicted areas. The most ambitious universities might aim for the carbon removal XPRIZE. Others might support restorative research, which enters the field to build back damaged or transformed ecosystems, like a University of Virginia effort to restore and expand seagrass on the Atlantic coast.[29] Higher education is already playing a research (and sometimes development) role for the climate crisis in the broader world. Universities could become the leading laboratories for humanity's Anthropocene transformation.

A slice of academic research turns its gaze inward to examine how academia responds to the climate crisis. We cited some examples in chapter 2, such as a political economy analysis of university endowment divestment campaigns. Elsewhere, there are philosophers who examine scholars in other disciplines, analyzing how they think through the crisis.

A Royal Melbourne Institute of Technology engineering team is studying colleges and universities as they plan for the climate crisis. Walter Leal, chair of Climate Change Management at the Hamburg University of Applied Sciences, has done tremendous work organizing conferences and scholarly collections on this topic.[30]

Institutional support for academic climate crisis research of all sorts may take various forms. We could see climate research centers appear, like Tsinghua University's Carbon Neutrality Research Institute, Oxford University's Environmental Change Institute, Johns Hopkins University's Ralph S. O'Connor Sustainable Energy Institute, or Rutgers University's Geoengineering Model Intercomparison Project. Some academic centers may consist of partnerships with local communities, such as the Centre for Climate Repair at Cambridge. Partnering with think tanks, like the Potsdam Institute for Climate Impact Research (*Potsdam-Institut für Klimafolgenforschung*) might be another way forward. Libraries can continue their historic and present role in supporting scholarly inquiry. A drive toward open access publications can make that role even more important. Libraries could also support the growing sharing and use of datasets, which might lead to new formats, genres, and publications.[31]

A great deal of this research may proceed through more partnerships with other academics and researchers from other institutions. For example, the World Weather Attribution project mentioned in chapter 5 involves researchers from the Red Cross Red Crescent Climate Centre, the Royal Netherlands Meteorological Institute, and the University of Oxford's Environmental Change Institute. Other research can include governmental scientists, as did a 2021 analysis of models seeking to account for land use in climate mitigation plans. Similarly, NASA's Jet Propulsion Laboratory joined forces with a nonprofit to develop a GHG identification tool. That nonprofit, Carbon Mapper, features scientists from academia, industry, and government. Four American colleges joined forces to buy electrical power from a distant solar farm. As part of that project, "students will intern with the solar farm developers, access the energy production data for research."[32]

Collaborations between academics and business may proliferate, as we saw in chapter 5 with Aarhus University working on wind turbine blade recycling with a group of collaborators, including turbine manufacturer "Vestas; chemical producer Olin, which produces resin for turbine blades [and] the Danish Technological Institute, an independent research and technology institute." Citizen science may enhance academic climate research, providing additional perspectives on and person-hours for projects. ClimatePrediction.net offers one example, where Oxford University scientists make available software that runs on thousands of other people's computers, distributively crunching data. The relationship between scholars and publishers could change as open access output and the amount of research data both grow, potentially driving new publication formats and genres.[33]

In pursuit of climate research, marginalized faculty will often need extra support. Untenured researchers will face multiple risks from external politics as well as on-campus forces. Faculty marginalized by gender, race, religion, sexuality, class, geography, and other axes of oppression, depending on social context, will need additional help in fending off bigotry and bias as they work. Antidiscrimination efforts that have been established recently may need to expand on this point, drawing on the emerging climate justice movement.[34]

Redesigning Teaching for the Emerging World

Turning from academic research to academic teaching, we can now explore how colleges and universities can prepare for changes to their educational mission, starting with certain fields facing physical challenges due to the crisis. For example, teaching agriculture may require a significant investment in that department's physical resources, which have deteriorated in some nations. We can think as well of classes that use remote sensor data, from engineering to environmental studies and oceanography. They may need backup plans as the environment changes.[35]

We should also prepare for climate change teaching that cuts across disciplinary divides. As noted earlier, the interdisciplinary nature of climate study may require consideration as a new form of the liberal arts. Alternatively, we could see teaching the Anthropocene taking the form of wide-ranging climate literacy, instantiated in many classes and through on-campus events. Such an initiative could link to the decades-long field of information literacy, teaching students how to detect climate disinformation and greenwashing. It can also become an emergent current within general undergraduate education. As University of Vermont engineering professor and university provost David Rosowsky asked his institution in 2016:

> What if we were the first major university to require all of our undergraduate students to have a minor or certificate in Climate? Why not? After all, professionals in every discipline will be required to come together to address the complex issues around climate change, and all of us will live in a world impacted by these changes. And why not UVM? We have nationally recognized faculty and programs in the environment and natural resources, climate studies, energy, water, complex systems, global and population health, policy, sustainability, food systems and agriculture, political science, and more.[36]

Academia may also teach itself, as it were. Our best understanding of the climate crisis may not be well understood by faculty and staff. A form of self-education or professional development may prove essential in order for the entire academic community to participate in thinking and planning for the next generations. Tecnológico de Monterrey calls this out openly in its new strategic plan: "Our commitment is to train the students and professors of Tecnológico de Monterrey to understand the complexity of the climate crisis and transform the world toward sustainable development."[37]

Higher education may also change the schedule of course offerings as the unfolding climate crisis reshapes the academic calendar. The

COVID-19 pandemic might provide one direction, as Beloit College shifted its semester schedule to a shorter one, based on "blocks," in order to provide more scheduling flexibility in case of threats or disasters. Alternatively, postsecondary institutions may reduce or end class offerings during the most dangerous times of the year, such as summers for areas experiencing intensified heat and humidity, or any month when dangerous storms are locally most likely to strike, such as late summer and fall for the Atlantic Ocean's hurricane season. On a shorter time frame, campuses in warmer climates, especially those with high humidity, may consider declaring "heat days" when rising temperatures spike to dangerous levels or shift a day's schedule to avoid holding classes and other meetings during the hottest hours.[38]

How we teach will probably change as much as *what* we teach. We can already identify a set of teaching practices that support climate learning, as we saw in chapter 3. Learning based on projects, student inquiry, and physical places are established pedagogies, practiced and studied, which we may expect to become more popular as the crisis demands more teaching. Service learning and internships are similarly known and may become more climate focused, with students working on storm damage cleanup or afforestation. Other less well-known pedagogies may surface and develop, driven directly or indirectly by the climate crisis. For example, peeragogy, the practice of students teaching one another without a guiding instructor, could grow among learners faced with disinterested or climate denier faculty. The movement to recover and honor Indigenous ways of knowing could inform postsecondary teaching. Given the globally disaggregated nature of higher education pedagogy, perhaps we should expect multiple schools of practice to claim the mantle of "climate teaching." We should also anticipate short, public teach-ins on climate topics, at times driven by current events, such as a disastrous storm or the implementation of a major policy.[39]

Digital learning may also rise, both in the form of wholly online education as well as computational work in face-to-face classes. Encouraging and supporting all of it requires some staff assistance and institutional commitment, ranging from an institutional library making available recording spaces to fully staffing an online teaching center. Off campus, there are many resources, from peers at other institutions to published research, experts on social media, and professional organizations. We should see a body of knowledge grow that describes digital climate teaching. The online community dedicated to helping people tell their climate crisis stories, hosted by StoryCenter, gives us an inkling about that possibility. StoryCenter takes care to support the stories of marginalized populations, offering a good example of a climate justice approach.[40]

The COVID-19 experience gives us further insights into how digital teaching might change. The most evident lesson is that higher education rapidly expanded its ability to teach entirely online. To the extent that weather damage impacts an academic institution and its population, this proven capacity may well be summoned throughout the rest of this century. For nations and regions suffering enrollment drops or financial stress, it seems that online teaching was a way to survive and thrive during the pandemic. On the other hand, the pandemic showed opposition to online learning. Dislike of "Zoom U" from faculty, students, and the popular imagination never fully faded, even as faculty and support staff scrambled to improve at times jury-rigged, emergency remote instruction. Our investment in in-person learning runs very deep, and perhaps not even the Anthropocene will quash it. COVID also coincided with expanding criticisms of Silicon Valley in general and ed tech in particular. The work of skeptics and critics such as Audrey Watters won an audience. Stories about proctoring companies and others violating student privacy and otherwise behaving as bad corporate citizens further tarnished the educational technology field. This animus could combine with concerns about campus technology's GHG footprint to drive some institutions to do less, not more, in digital education.[41]

At another level, we can think (or rethink) of teaching in terms of creative, cross-disciplinary problem study and solving. The climate crisis does not fit neatly into university curricular pigeonholes, nor does it tamely follow our intellectual domains. Instead, it remorselessly attacks our world across those conceptual lines. Teaching our students how to grapple with the crisis on its terms means taking seriously real-world learning and learning how to learn in a fluid situation, how to innovate in unfriendly environments, and how to communicate in unsupportive situations. Students need to learn adaptation, much as their colleges and universities do.

We should anticipate changes to academia's educational mission beyond pedagogy and the curriculum, starting with care for the student population. As demand for psychological counseling services grows in the world at large, counseling may also become more important on campus. We may anticipate increasing interest in providing or referring students to mental health services as eco-anxiety rises and climate-caused traumas burgeon, as noted in chapter 3. This is not just an American phenomenon, despite US colleges' and universities' high standard of student care. A recent *Lancet* article called for a broad investment in psychological support, including for "training for health professionals on climate change and mental health [to] be increased." The global activist group 350.org calls on the world to prepare for climate grief; academia should not be exempt from this. Such preparation could take the form of hosting, or allowing members of the community to host, climate cafes. Modeled on death cafes, these are social spaces for people to safely express their feelings about this fraught topic in a social, facilitated setting. On the biological health side of care, we should plan for changes to what physical health care a campus offers as conditions alter. For example, a campus on the edge of desertification should expect to see worsening cases of asthma. A college susceptible to flooding can plan on more cases of mold allergies. Institutions likely to be hit by weather damage in the short term should expect a wide range of injuries. Institutions should take care to support their students only after learning more about them, without presumption. Population surveys and focus groups may be a good way to start.[42]

Higher education offers other learning services apart from classes, which the climate crisis can alter. "Work colleges," institutions that require students to perform labor with their hands beyond class assignments, may offer manual labor options specifically concerning the climate crisis, such as repairing levees or preparing food for refugees.

Career services will be a focal point for directing students to "green jobs," working in decarbonization enterprises and post-fossil fuel industries. Career services can also guide graduating students to positions that help humanity through the crisis: disaster management, government, certain nonprofits, and the like. Student groups can organize along similar lines, linking students to jobs as well as political action. Student journalists can conduct climate research with strong local resonance. The University of Florida student newspaper offered an interesting foretaste of this during the COVID pandemic, when its rising journalists critically dug into their campus's infection data.[43]

The experience of students living on a campus may change in some ways. Residential universities and colleges may host alternative living spaces with a climate theme, such as the Domes at the University of California, Davis.[44] Residence life can offer climate crisis programming. At the same time, residential institutions may rethink the climate footprint of their activities, such as travel-intensive spring break. They might reduce or remove water-based entertainment (swimming pools, lazy rivers) in arid areas.

Teaching off campus may also rise, and with some urgency if an institution deems the crisis of such importance. As noted in chapter 3, academic libraries and museums can present interactive climate change content and experiences to the community and travelers in a form of public outreach, somewhat like that offered by museums. This can be a way of expanding access to a college's or university's intellectual resources, as well as trying to better inform the public along climate lines. Nonacademic museums are already doing this. For example, the Peabody Essex Museum has recently offered a series of exhibits and programs about the climate crisis. Notably, they are aimed at getting visitors to think productively about climate futures.[45]

Such outreach brings us to the topic of academic-community relations, which we explored in chapter 4. Colleges and universities have an additional strategic path to follow—namely, improving community relations through work on climate issues. There are many projects and avenues to explore, as we have seen. In addition to that discussion, we could add other examples of potential town-gown collaboration. We could also see students, faculty, and staff working on local biological matters, such as proactively planting species most likely to thrive in the emerging climate while transporting away those that will not succeed, altering landscapes in two or more communities. In countries where police reform has become a vibrant issue, academics could work with local parties on connections between climate justice and police reform. As activist Sam Grant put it, some populations are more at risk than others from both climate *and* police violence. Some police forces politically align with oil companies and their infrastructure, interacting with populations accordingly. Moreover, there's a question of risk in political organization along these lines: "The fight for climate integrity necessitates that all human beings who care about their future are going to be standing up, practicing democracy. And if as we stand up now, we're going to have our lives put at risk by law enforcement who are trained and paid to protect property at all expense at any cost."

What is the responsibility for non-marginalized academics to support the marginalized in climate activism?[46]

The Potential of Interinstitutional Collaboration

So far, we have spoken of individual colleges and universities, yet there is so much potential in collaborations among multiple institutions. As we advance further into the developing climate crisis, campuses can learn from one another's experiments and practice. Campuses can also build shared resources, exchange classes, and mutually support professional development networks. There are already examples of interinstitutional collaboration occurring around climate change. International Universities Climate Alliance is a scholarly information-sharing network of nearly fifty research universities on six continents. The Assessments of Impacts and Adaptations to Climate Change (AIACC) initiative linked a number of sub-Saharan universities to build their teaching capacity. Latvia's Council of Environmental Science and Education (LVZIP) helped organized that nation's universities and external experts in developing on-campus and public education. In 2009, vice chancellors of five universities within the Australian Technology Network (ATN) formed a collaborative Emission Reduction Working Group to lower their collective GHG output. Second Nature works at a presidential level to bring together institutions aiming to reduce carbon emissions, including facilitating offset purchasing. The Cooperative Institute for Climate, Ocean, and Ecosystem Studies (CICOES) links Oregon State University, the University of Washington, and the University of Alaska Fairbanks, as well as the National Oceanic and Atmospheric Administration in collaborative research efforts. The EcoLeague brings together small, liberal arts colleges for a student exchange program.[47]

Some collaborations work through facilitating third parties, like the Sustainable Endowments Institute, which provides tools and competitions for participating institutions, or the Adaptation Learning Network, which provides learning resources to academics in British Columbia. The Virginia Foundation for Independent Colleges helped create the Collaborative Heat Mapping Research Project, wherein students measure temperatures of local urban areas in detail. We may see more of these collaborations, and more ambitious ones, as the climate crisis progresses.[48]

Beyond the academic world, colleges and universities can play a role in the Anthropocenic world, as we saw in chapters 4 and 5. We can see more options as we look ahead. Faculty and staff can enter public debates through various means (old media, social media, campus outreach, personal networks) and seek to inform and influence discussions and decisions. This already occurs, of course, with examples easy to find, like a multilingual, international group of scientists in 2021 publicly urging the world to cut down on methane emissions. A 2021 National Academy of Sciences report authored by nearly one hundred scientists intervened in the geoengineering debate, putting forth a model for broad-based public input. In 2020, another group of scientists called for global action and included academic researchers in their urgency: "As the Alliance of World Scientists, we stand ready to assist decision-makers in a just transition to a sustainable and equitable future. We urge widespread use of vital signs, which will better allow policymakers, the private sector, and the public to understand the magnitude of this crisis, track progress, and realign priorities for alleviating climate change."[49]

Institutions can support their faculty (and staff, and some students) as they act in such public intellectual roles. Such support can take the form of media outreach, media production, and backing up academic freedom.

In 2019, a world group of universities and colleges signed off on a letter urging world decarbonization, set in the framework of the United Nations' Sustainable Development Goals (SDGs). This suggests one way academics can proceed in their climate actions, by connecting their work to those seventeen goals. Each goal gives a clear way to communicate and connect with global audiences, including governments, NGOs, funders, and civil society, especially when thinking of justice and equity. A university could, for example, cast its effort to source electrical power entirely from local solar and wind in terms of SDG number 7, Affordable and Clean Energy.[50]

Longer Term

Resilience asks us "how do we keep what we really want to keep?" Relinquishment asks us "what do we need to let go of in order to not make matters worse?" Restoration asks us "what can we bring back to help us with the coming difficulties and tragedies?" Reconciliation asks "with what and whom can we make peace as we face our mutual mortality?"

—JEM BENDELL[51]

The first half of this chapter describes what academia can do now and in the short term. These options are based on realizing effects within short- and medium-term horizons. If we advance those strategic horizons further on, to a much longer term, to the rest of the century, what can colleges and universities start doing or considering now? What can we best contribute to the next two generations? Or do we have the capacity to think, plan, and act along such timelines? If we can speak of students taking a sixty-year curriculum, can academia think, plan, and act on a span of eighty years?

Consider as an example the possibility of long-term biological preservation. Postsecondary institutions could set up refuges for flora and fauna, either protecting them against a changing climate or welcoming them to a new habitat. We could see a campus conducting such efforts as following in the footsteps of the Svalbard Global Seed Vault. As an early example of this idea, one California public university recently started supporting "a menagerie of animals kept at UC Davis until their environments become cool enough to live in again."[52] The quote refers to an unusually high fire and temperature season, but that obviously stems from global warming. Perhaps other universities and colleges around the world will attempt such long-term acts of preservation. Conceptually, such actions would echo literary visions of preserving the present against future catastrophes, à la Isaac Asimov's *Foundation* (1951).

Similarly, institutions can work with long-term climate mitigation or adaptation projects hosted and directed beyond the academy. As of this writing, there are now two massive efforts on two different continents with closely related names and concepts, both aimed at blocking desertification. In China, the Three-North Shelter Forest Program (三北防护林) builds massive forest lines to stop the Gobi Desert from advancing farther into China. Unsurprisingly, "Great Green Wall" is a popular nickname. In sub-Saharan Africa, an African Union project, the Great Green Wall of the Sahara and the Sahel (Grande Muraille Verte pour le Sahara et le Sahel), does the same against that continent's immense desert, as noted in chapter 6. These are not multiyear but multi-decade projects. What role can colleges and universities play with such efforts? Could we consider donating campus-owned land to such efforts, or also to supporting public or private conservancy efforts? Can we contribute our research to refine them while researching their progress? Should we teach them? Should students, along with staff and faculty, contribute labor? How can academic institutions match their planning cycles to projects that plan on generational scales?

Thinking about the climate crisis and academia in the long term also allows us to anticipate fortuitous moments when separate decarbonizing forces reinforce one another. Colleges and universities should consider benefitting from, and participating in, what Robinson Meyer calls "the green vortex," the process by which many civil society actors cocreate new post-carbon systems and dynamics in a bottom-up fashion: "how policy, technology, business, and politics can all work together, lowering the cost of zero-carbon energy, building pro-climate coalitions, and speeding up humanity's ability to decarbonize."

Scanning for such moments requires studying numerous domains and projects. Few organizations on Earth are as well equipped to perform such analysis as colleges and universities.[53]

If these long-term ideas are daunting, it helps to recall that some of higher education already works on that scale through smaller steps. Some short-term changes accumulate and become more significant in a series over the long term. Small alterations and additions to a campus's built environment gradually build up to a larger transformation. We can stroll through the world's oldest universities and see an architectural history on display, as buildings appear from different eras and styles. Accordingly, we can imagine what a college or university looks like by, say, 2060, once buildings carry gardens, more structures are built from wood and wood-based materials, more solar panels appear, and more DAC machines are present. Perhaps we will be able to identify faculty offices built from wood composites in the Berlin-Kreuzberg "WoHo" style from the early 2030s, while walking on a solar cell–embedded path laid down in the 2020s.[54] Or consider how a college or university would appear after restricting or banning fossil fuel–powered vehicles on campus. That alters the look and feel of an institution, especially as roads and parking spaces change or disappear. Step by step, something perhaps like a solarpunk vision replaces what we now see in our contemporary institutions. Or some new styles will appear. We can envision them now.

Another aspect of modern academia follows this incremental change driving major transformation: the education of women, who now constitute a majority of students in some nations for the first time in history, following decades of steady, gradual enrollment growth. This has had many benefits for those women and their societies in general, and may also have a powerful impact on the climate crisis if postsecondary higher education continues the practice. For those concerned that high population levels worsen global warming by adding to the number of consumers burning carbon, we know that more education correlates with women having fewer children. As projections show more women than men suffering from climate change's stresses, having more education gives women more resources for survival. As a Rapid Transition Alliance report put it:

> Girls who completed their schooling are more likely to be able to use their knowledge and leadership to support their families during climate shocks by finding better-paying jobs. An educated girl is simply better equipped with the skills to withstand and overcome the shocks of extreme weather events and changing weather cycles. A 2017 Brookings [Institution] study suggests that for every additional year of schooling a girl receives on average, her country's resilience to climate disasters can be expected to improve by 3.2 points (as measured by the ND-GAIN Index, which calculates a country's vulnerability to climate change in relation to its resilience).

Drawdown (2017) is a popular and optimistic collection of climate strategies, which we've cited several times so far. It concludes that among all of the strategies it considers, from technologies and policies, the most effective is educating women. Educating more women is one powerful way for colleges and universities to help the human race better respond to the climate crisis along the timeline of multiple human lifespans.[55]

Classically, the role postsecondary institutions play in educating traditional-age (eighteen to twenty-two years old) young people has long had the potential to transform society over the course of generations. Right now, some passionate climate activists seek to set their campus aflame with protests, strikes, and other actions. Their experience in these institutions will help shape where they try to take the world. And non-activists also pass through academia's gates. For several years, higher education gets to help shape their thinking. When it comes to climate change, these institutions actively form the future of tens of millions of people every few years.[56]

What we teach will vary over time, even through the Anthropocene. Looking ahead over decades, academia should anticipate rising interest in, and support for, curricula people perceive as most directly addressing the climate crisis. As noted in chapter 3, we can view this as a supply and demand problem. Rising student interest in climate classes can cause colleges and universities to supply more of them, depending on an institution's structure, resources, and culture. Faculty members, academic programs, and those publishing content can supply more materials and opportunities for learning. As with any supply and demand model, the amounts of each can fluctuate over time, depending on circumstances. Major climate events, such as disasters occurring or megaprojects in progress, may drive demand up. In contrast, a declining sense of urgency could depress the number of students taking an Advanced Climate Mitigation seminar. Too much supply can also outpace demand. It would be too simple and inaccurate to say, "The Anthropocene will drive up related classes for two generations." It is more accurate and practical to think instead of paying careful attention to signs of interest and to anticipate their next contours.

For example, a consensus is emerging that a clear path lies ahead: switching from fossil fuels to electrical engines powered by renewables. If enough of society accepts this premise and works to realize it practically, then we may see enormous demand for workers who can make that switch in the energy sector—which means growing demand for providing students with those skills, from electrical engineering on. Anticipating this will take some academic foresight. Following up on it with curricular changes, faculty and support staff hires, on-campus infrastructure, and so on will require political and financial capital. How many institutions are thinking along these lines? Yet how many are planning past that point? If the human race successfully switches off carbon and powers civilization by electricity backed by solar, wind, and others, just how badly will demand for those jobs decline? Again, looking ahead for nearly eighty years entails being sensitive to, and anticipating, such changes in demand and supply curves.

Universities in New Economies

Thinking long term necessitates forecasting economic developments and planning strategic responses to them. As we've seen throughout this book, the climate crisis will cause economic damage around the world, albeit in varying ways depending on local circumstances. Colleges and universities would do well to anticipate such hits to their budgets. Further, economic inequality has increased in many nations, and class divides seem likely to widen. The COVID-19 pandemic accentuated this. Such a changing economy gives higher education several challenges, including having to rethink career services and job placement. It may also mean that colleges and universities will find it more practical to approach the very wealthy for assistance along climate change lines than they currently do, especially in nations that have privatized higher education to significant degrees. We have already seen the superrich play roles in trying to mitigate or adapt to global warming. To pick one example, Amazon founder Jeff Bezos launched the Bezos Earth Fund, a "$10 billion commitment to fund scientists, activists, NGOs, and others." Higher education may see fit to work with billionaires like Bezos to support academic work. In addition, given the immensely powerful role capital plays in shaping the world, engaging with the financial industry may be a way to better grasp and influence our collective response to the crisis.[57]

We may also anticipate an economic future that differs from the post–Cold War experience. As noted in chapter 5, some regions and nations may respond to the Anthropocene by planning to halve economic growth or to drive for an economic build-down. As two physicists have argued through modeling, "Degrowth would not be an easy solution, but, as indicated by our results, it would substantially minimise many key risks for feasibility and sustainability compared with established, technology-driven pathways." If the region or nation where a campus is located pursues such a major shift in its economic structure, what does this mean for the institution? Universities depending on financial growth from investments or donors may see those supports flattened. Can academic economists play a role in no-growth or degrowth planning now? What light can other disciplines, such as sociology or psychology, shed on such an evolving scenario? How would such futures appear in our curricula?[58]

Populations and Some Campuses in Flight

Planning and enacting such economic shifts will become more difficult as climate migration grows. As noted earlier, the number of climate refugees should expand over the next generations as environmental pressures and disasters hit certain areas with increasing frequency and ferocity. Agricultural declines, droughts, spiking wet-bulb temperatures, and storms will drive more people into migration, building up to tens or *hundreds* of millions of refugees by 2100, according to the World Bank. Political responses to these problems—nationalization of resources, militarization, civil unrest, civil war, interstate conflict—can add even more numbers to this rising nomadic tide. Political responses to the burgeoning refugee movement may demonstrate inhumanity, as the twenty-first century has amply demonstrated, which can worsen the problem while also creating additional problems, including more public forms of bigotry and more strident nationalisms.[59]

Let us now number some academic institutions in those danger zones. How will the academy respond to such a threat, looking ahead two generations? To begin with, a substantial number of colleges and universities are presently located in danger zones, notably the warmest areas and those most exposed to violent weather. Campuses near oceans run the risk of flooding or total immersion. Those will have to decide on a survival or adaptation strategy. They may seek shelter behind a seawall, elevate buildings on stilts and piles, raise their grounds above projected oceanic heights, migrate to another location, or simply come to an end. Institutions near growing deserts should pursue other measures, including building defensive forests or erecting anti-sand barriers. Those institutions will face dire problems, such as how to attract people to their grounds when the local community drains away. How can they afford to protect themselves? That is, how much air-conditioning can a campus budget? How often can its financial resources handle repairs from weather damage and rising insurance costs? How many lawsuits over personal injury and death can a school handle? If the local area does not mount general protective measures against floods, what kind of elevation or wall can a university afford? Questions are open as to how much assistance governments, nonprofits, the rich, crowdfunding, and corporations may provide, and with what conditions. It seems that many institutions located in danger areas will be faced with the real choice of closure or relocation. Governing bodies and various stakeholders will have ideas about how this should proceed. Such foundational, radical, and existential questions are what the long view entails.

Campuses located in areas free of such imminent dangers will face other choices as they decide how to respond to destruction suffered elsewhere and refugees therefrom. First, what calculations will go into deciding whether or not to host refugees on site? What pressures will influence those choices, both from within the institution and from beyond? Some universities and nonprofits have urged colleagues and governments to expand their support in 2021. Will this become a mainstream behavior of campuses? Second, what academic support can higher education provide to migrants? We've already discussed some of the issues here, including face-to-face versus online instruction. Campuses could offer scholarships aimed at climate refugees, as Moses Seitler has urged. We could imagine academic programs or entire colleges devoted to educating and supporting climate refugees. Overall, an ambitious or well-resourced campus could position itself as a climate sanctuary. Its communication strategy could celebrate this mission or minimize it, of course, because of an open question that would attend that development: how being an academic climate sanctuary changes an institution's reputation. Departments, programs, colleges, universities, and systems in the global north, and unthreatened with immediate climate disaster, may see supporting refugees from the global south as an act of large-scale climate justice or reparations.[60]

Another question will appear more often: what responsibilities will colleges and universities in safe areas have to those colleagues under attack? Should academic units take steps to provide resources to their peers, from professional development to scholarly support? How many will offer to help teach out (i.e., offer students from a closing institution classes to finish their degrees) students enrolled in institutions shutting down or suspending operations for a relocation? Will the climate refugee crisis give extra power to calls for open educational resources (OER) and open access (OA) scholarship in order to remove one barrier to learning for those suffering from involuntary migration? We could imagine regional or national shifts as endangered institutions gradually shift people and services toward those in safer areas. This could lead to a reduction in the number of campuses while increasing the size of remaining institutions. Ultimately, the climate crisis could drive the expansion of mega-universities.

The preceding paragraphs distinguished between areas subject to climate devastation and those that are not, but that opposition might become less clear as the decades unroll and the Anthropocene deepens. For example, central Russia or the middle of the United States does not presently enjoy seafronts nor include areas with excessive heat, but both of these regions may gradually lose stocks of potable water as demand for it increases and supplies are drawn off.[61] In addition, I specified "zones" and "regions," rather than nations, because academia within a given country could experience both situations. Multiple schools within a given system, be it a province, church, or state, could receive natural crises in very different ways.

Beyond the direct threats posed by the climate itself are the secondary and tertiary effects discussed in the introduction to this book, those stemming from the natural and human worlds. For the latter, roiled polities and economies present massive challenges to the higher education sector for the rest of the century. How prepared are campuses today for an economic recession or collapse in the next decades? With what foresight is the academic world planning on civil unrest, revolution, or war? How will campuses grapple with cultural ferment, from new religious movements to insurgent groups trying to reshape society during the worsening Anthropocene? If we can imagine a university as a climate sanctuary, just how far is that institution willing to go during a civil war?

Here I would like to return to chapter 6's worst-case scenario and pick up my earlier reference to Asimov's *Foundation*. One function higher education may see itself performing over the next two to three generations is preserving knowledge and attitudes during a general collapse. If governments are falling amid growing starvation, mass migrations, terrible storms, and social upheaval, a campus may seek to hold candles against that darkness. Preserving knowledge is, after all, a longstanding and core feature of the university. We might see such institutions erect ever higher barriers against those literal and figurative rising tides, protecting their precious intellectual stock. Or we could see multiple institutions collaborating in distributing their academic resources in the internet's original spirit. This may sound hyperbolic or fantastic to some readers, yet the prospects of such a disastrous century are neither.

Perhaps the academy will take more proactive steps over the following decades. There is a tradition of activism starting from within universities and colleges. Student movements stud the historical record, including the modern environmental movement. Campuses may incubate organizers and activists who move on to participate in the great climate struggle. Some academic research will appear within public discourse and play a role in decision-making from local communities to the international community. So much of this will probably occur without academia taking conscious steps to make it so, simply because of our somewhat decentralized nature, when broadly compared to the rest of society, and to the connections with the crisis this book has outlined so far.

Could higher education go further?

We may imagine chancellors and faculty senates declaring institutional commitments to climate mitigation and adaptation, then putting resources behind such strategies. These colleges and universities would cease to position themselves as neutral parties and instead intervene directly in the world, even at a small scale. They could bring to bear their full gamut of intellectual capability and also run the risk of opposition. They would teach what Peter Sutoris calls "Anthropocene skills," combined scientific and political abilities that give students ways of participating productively in the warming world, then follow those students in intervening within the crisis.[62]

Campuses can also organize together for this purpose and do so generatively. We have already seen institutional leaders agreeing on the importance of lowering carbon emissions. Imagine them forming leagues with specific programs and missions. A group of colleges might agree to end long-haul travel from its members, or collaboratively invest in purchasing DAC hardware. We could see an academic alliance creating an international cadre of students committed to studying and conducting fieldwork to mitigate and adapt to the crisis, a kind of honors society or academic order to which it will be an honor to belong. Obviously, this would take serious work to bring about and support, and just as obviously would run into risks and opposition. How many academics would find the cause worth the effort?

Trent Batson has envisioned something along these lines with the Last Humans Project. This is a call for colleges and universities to align their mission statements and curricula in order to teach a new generation Peter Sutoris's Anthropocene skills through a mix of intellectual and hands-on work, not only for their personal benefit but to try reshaping humanity's climate strategy. The goal would be to influence civilization as a whole, to "lead the world in preserving human society during the current mass extinction caused by climate change." Batson envisions a massive movement or cadre educated and supported by higher education: "If only one tenth of [currently enrolled] students [worldwide] opt for our special "extinction rebellion" curriculum of active learning in real-world extinction projects, The Last Humans Project will have a very large Survival Force (25,000,000) to join the Project and help preserve human society."[63]

How many academics around the world would see this as a powerful and proper use of our position in the world? How many might change their minds over the next several decades as the crisis proceeds? To ask the question another way, how many would see *not doing so* as an act of abdication, of turning away from humanity in the teeth of a global disaster?

In thinking so ambitiously, we can return to a theme we've hit on throughout this book. Will academic institutions take these steps and make these plans inclusively? The history of universities offers an appalling body of evidence demonstrating our ability to exclude people based on nationality, gender, religion, class, geography, and more. Instead, our institutions must advance into the climate crisis by including populations seriously and with care. Rethinking our curricula and research enterprises should not reproduce inherited practices of bias and oppression. Working with the world, from local communities to the international community, must not reprise colonialism.[64]

Within our institutions, will plans and actions engage all people and groups? Once again, the global history of academia gives ample evidence of institutional management practices that involve only segments or even slivers of a university or college community. We also have traditions of community involvement and campus democracy. Will we grapple with the climate crisis in such a participatory fashion, or rerun our bureaucratic and mandarin practices?

I would like to conclude on a note of intense urgency. The climate crisis may be the greatest threat colleges and universities face in this already volatile, uncertain, chaotic, and ambiguous century. Global warming threatens to squeeze, injure, and destroy institutions and the many purposes we invest in them. Yet at the same time as academia faces this challenge, we also have the opportunity to improve humanity's ability to understand, mitigate, and adapt to the emergency. I hope this book has shown that higher education has the capacity to contribute in a major way through research, teaching, community relations, and public activism.

I return to the words of the Bulletin of the Atomic Scientists, which we first considered in the introduction to this book, when that organization surveyed the world's threat landscape in 2021 and determined that humanity was botching the job: "The [COVID-19] pandemic serves as a historic wake-up call, a vivid illustration that national governments and international organizations are unprepared to manage nuclear weapons and climate change, which currently pose existential threats to humanity."

If we were this unprepared to meet challenges in 2021, our danger will be that much greater as the climate crisis ratchets up remorselessly over the years and decades to come. For those national governments and international organizations, academia could play a vital role in helping them become better informed, wiser, and more resilient. That might not be a gentle process. Higher education may have to coerce, cajole, persuade, and sometimes shock the global system into resilience through our social connections, intellectual firepower, and ability to nurture.

Yet it is too late now for colleges and universities to begin preparations for a far-off danger. The crisis is already upon us. We are advancing ruthlessly into the Anthropocene. Fires now burn on academia's horizon. It is up to us to choose if those will be flames of destruction or the lights of illumination.

NOTES

Epigraphs

Elizabeth Kolbert, *Under a White Sky: The Nature of the Future* (New York: Crown, 2021), 94. He Pou a Rangi New Zealand Climate Change Commission, *Ināia Tonu Nei: A Low Emissions Future for Aotearoa* (Wellington: He Pou a Rangi Climate Change Commission, 2021), https://ccc-production-media.s3.ap-southeast-2.amazonaws.com/public/Inaia-tonu-nei-a-low-emissions-future-for-Aotearoa/Inaia-tonu-nei-a-low-emissions-future-for-Aotearoa.pdf. Lonnie G. Thompson, "Climate Change: The Evidence and Our Options," *Behavior Analyst* 33, no. 2 (Fall 2010): 153–70.

Introduction. Academia Wades into the Anthropocene

Epigraph. Kim Stanley Robinson, *The Ministry for the Future* (New York: Orbit), 2020.

1. World Higher Education Database, https://www.whed.net/home.php; "How to Claim a Place amongst the Top 1% of World Universities?," *QS* (blog), June 9, 2017, https://www.qs.com/claim-place-amongst-top-1-world-universities/; Angel J. Calderon, "Massification of Higher Education Revisited," Royal Melbourne Institute of Technology, 2018, http://cdn02.pucp.education/academico/2018/08/23165810/na_mass_revis_230818.pdf; Richard Price, "The Number of Academics and Graduate Students in the World," *Richard Price* (blog), November 15, 2011, https://www.richardprice.io/post/12855561694/the-number-of-academics-and-graduate-students-in; Joshua Kim, "The $1.9 Trillion Global Higher Ed Market," Inside Higher Ed, May 2, 2017, https://www.insidehighered.com/blogs/technology-and-learning/19-trillion-global-higher-ed-market.
2. Robert Cushman (Pacific Union College president) and Maria Rankin-Brown (academic dean), interview by the author, 2020; Tiara Walters, "'Out-of-Control' Table Mountain Fire Forces UCT Evacuation," *Daily Maverick* (Johannesburg), April 18, 2021; Harry Cockburn, "Extinction Rebellion and Greenpeace Back Boycott of Science Museum's Shell Sponsored Climate Exhibition," *Independent*, April 28, 2021, https://www.independent.co.uk/climate-change/news/science-museum-shell-boycott-petition-b1838958.html; Internal Displacement Monitoring Centre, *Global Report on Internal Displacement: Internal Displacement in a Changing Climate* (Geneva: Internal Displacement Monitoring Centre, 2021), https://www.internal-displacement.org/sites/default/files/publications/documents/grid2021_idmc.pdf; Sandra Steingraber, "Commentary: A Farewell to Ithaca College after 18 years," *Ithacan*, March 3, 2021, https://theithacan.org/opinion/commentary-a

-farewell-to-ithaca-college-after-18-years/; Oliver Milman, “US Climate Research Outpost Abandoned over Fears It Will Fall into Sea,” *Guardian*, April 14, 2021, https://www.theguardian.com/us-news/2021/apr/14/cape-cod-climate-research-station; Bryan Alexander, *Academia Next: The Futures of Higher Education* (Baltimore: Johns Hopkins University Press, 2020), 182–88, 196–201.

3. Jose Guillermo Cedeño Laurent, Augusta Williams, Youssef Oulhote, Antonella Zanobetti, Joseph G. Allen, and John D. Spengler, “Reduced Cognitive Function during a Heat Wave among Residents of Non-Air-Conditioned Buildings: An Observational Study of Young Adults in the Summer of 2016,” *PLoS Medicine* 15, no. 7 (July 2018), https://doi.org/10.1371/journal.pmed.1002605; David Hasemyer, “Trouble in Alaska? Massive Oil Pipeline Is Threatened by Thawing Permafrost,” NBC News, July 11, 2021, https://www.nbcnews.com/news/us-news/trouble-alaska-massive-oil-pipeline-threatened-thawing-permafrost-n1273589.
4. Brian Fagan, *The Attacking Ocean: The Past, Present, and Future of Rising Sea Levels* (New York: Bloomsbury, 2013), 186.
5. “Climate Change and Health,” World Health Organization, October 30, 2021, https://www.who.int/news-room/fact-sheets/detail/climate-change-and-health.
6. Philip Alston, *Climate Change and Poverty: Report of the Special Rapporteur on Extreme Poverty and Human Rights* (Geneva: Human Rights Council Forty-First Session, June 24–July 12, 2019), https://digitallibrary.un.org/record/3810720?ln=en; Ryan D. Harp and Kristopher B. Karnauskas, “Global Warming to Increase Violent Crime in the United States,” *Environmental Research Letters* 15, no. 3 (March 2020), https://iopscience.iop.org/article/10.1088/1748-9326/ab6b37; Michelle Horton, “Stanford Researchers Explore the Effect of Climate Change on Suicide Rates,” *Stanford News*, March 29, 2019, https://news.stanford.edu/2019/03/29/effects-climate-change-suicide-rates/; Christian Parenti, *Tropic of Chaos: Climate Change and the New Geography of Violence* (New York: Bold Type Books, 2011).
7. Kathleen Fitzpatrick, *Generous Thinking: A Radical Approach to Saving the University* (Baltimore: Johns Hopkins University Press, 2021); Sara Goldrick-Rab, *Paying the Price: College Costs, Financial Aid, and the Betrayal of the American Dream* (Chicago: University of Chicago Press, 2016); “Generous Thinking with Kathleen Fitzpatrick,” Future Trends Forum, YouTube video, June 10, 2019, https://youtu.be/Kzu00eV7JJI.
8. Scripps Institution of Oceanography, *Climate Change and Sea Level Rise: Scenarios for California Vulnerability and Adaptation Assessment*, CEC-500-2012-008, (Sacramento: California Energy Commission, July 2012), https://uc-ciee.org/ciee-old/downloads/Climate%20and%20Sea%20Level%20Rise%20Scenarios%20for%20California.pdf; Fagan, *Attacking Ocean*, 12; Working Group II, *Climate Change 2022: Impacts, Adaptation and Vulnerability: Summary for Policymakers* (Geneva: Intergovernmental Panel on Climate Change, 2022), https://www.ipcc.ch/report/ar6/wg2/downloads/report/IPCC_AR6_WGII_SummaryForPolicymakers.pdf; Nicola Jones, “How the World Passed a Carbon Threshold and Why It Matters,” Yale Environment 360, January 26, 2017, https://e360.yale.edu/features/how-the-world-passed-a-carbon-threshold-400ppm-and-why-it-matters.

9. These are books I recommend to anyone looking to learn more about the climate emergency. Each has a different focus. David Wallace-Wells, *The Uninhabitable Earth: Life after Warming* (New York: Tim Duggan Books, 2020); Christiana Figueres and Tom Rivett-Carnac, *The Future We Choose: Surviving the Climate Crisis* (New York: Vintage, 2020); Elizabeth Kolbert, *The Sixth Extinction: An Unnatural History* (New York: Henry Holt, 2014); Amitav Ghosh, *The Great Derangement: Climate Change and the Unthinkable* (Chicago: University of Chicago Press, 2017); Naomi Klein, *This Changes Everything: Capitalism vs. the Climate* (New York: Simon and Schuster, 2015); Paul Hawken, *Drawdown: The Most Comprehensive Plan Ever Proposed to Reverse Global Warming* (New York: Penguin, 2017); Katharine K. Wilkinson and Ayana Elizabeth Johnson, eds., *All We Can Save: Truth, Courage, and Solutions for the Climate Crisis* (New York: Penguin, 2020); Michael E. Mann, *The New Climate War: The Fight to Take Back Our Planet* (New York: PublicAffairs, 2021). For film and video, I recommend *An Inconvenient Truth* (2006) and the much angrier *The Age of Stupid* (2009). David Attenborough's climate change work is excellent, including *Climate Change: The Facts* (2019) and *Breaking Boundaries: The Science of Our Planet* (2021).
10. Timothy Morton, *Hyperobjects: Philosophy and Ecology after the End of the World* (Minneapolis: University of Minnesota Press, 2013).
11. Jamais Cascio, "Facing the Age of Chaos," Medium, April 29, 2020, https://medium.com/@cascio/facing-the-age-of-chaos-b00687b1f51d.
12. Stephen Nash, *Virginia Climate Fever: How Global Warming Will Transform Our Cities, Shorelines, and Forests* (Charlottesville: University of Virginia Press, 2014).
13. Elizabeth Kolbert, *Under a White Sky: The Nature of the Future* (New York: Crown, 2021), 94; Bill McKibben, *Eaarth: Making a Life on a Tough New Planet* (New York: St. Martin's, 2011).
14. Roger L. Geiger, "The Ten Generations of American Higher Education," in *American Higher Education in the Twenty-First Century: Social, Political, and Economic Challenges* 4th ed., ed. Michael N. Bastedo, Philip G. Altbach, and Patricia J. Gumport (Baltimore: Johns Hopkins University Press, 2016).
15. Fiona Harvey, "Anxious about Climate: 4 in 10 Young People Are Wary of Having Kids," *Mother Jones*, September 15, 2021, https://www.motherjones.com/environment/2021/09/global-survey-climate-change-anxiety-young-people-children-kids/; Alexander, *Academia Next*.
16. Alexander, *Academia Next*, 23.
17. Yojana Sharma, "Disaster Preparedness Would Improve HE Pandemic Response," University World News, July 18, 2020, https://www.universityworldnews.com/post.php?story=20200715113545432/.
18. Bruno Latour, "La crise sanitaire incite à se préparer à la mutation climatique," *Le Monde*, March 25, 2020, https://www.lemonde.fr/idees/article/2020/03/25/la-crise-sanitaire-incite-a-se-preparer-a-la-mutation-climatique_6034312_3232.html; Christina Birch, Rodger Edwards, Sarah Mander, and Andy Sheppard, "Electrical Consumption in the Higher Education Sector, during the COVID-19 Shutdown," IEEE Power Engineering Society Conference and Exposition in Africa, 2020,

https://ieeexplore.ieee.org/document/9219901; Geoff Mann and Joel Wainwright, *Climate Leviathan: A Political Theory of Our Planetary Future* (New York: Verso, 2018).

19. Dominic Mealy, "To Halt Climate Change, We Need an Ecological Leninism: An Interview with Andreas Malm," *Jacobin*, June 15, 2020, https://www.jacobinmag.com/2020/06/andreas-malm-coronavirus-covid-climate-change.
20. John Mecklin, ed., "This Is Your COVID Wake-Up Call: It Is 100 Seconds to Midnight 2021, Doomsday Clock Statement," Bulletin of the Atomic Scientists, 2021, https://thebulletin.org/doomsday-clock/2021-doomsday-clock-statement/; "COVID-19 Dashboard by the Center for Systems Science and Engineering (CSSE) at Johns Hopkins University (JHU)," https://coronavirus.jhu.edu/map.html.
21. Bryan Alexander, "COVID-19 on Campus: A New York Times Survey and a Data Void," *Bryan Alexander* (blog), August 28, 2020, https://bryanalexander.org/coronavirus/covid-19-on-campus-a-new-york-times-survey-and-a-data-void/.
22. Harry Clarke-Ezzidio, "Why We Need to Talk about 'Global Weirding': Discussing Climate Change in Terms of 'Global Warming' Can Spark Confusion. Is There a Better Way to Describe Strange Weather Patterns?," *New Statesman*, April 20, 2021, updated July 23, 2021, https://www.newstatesman.com/science-tech/2021/04/why-we-need-talk-about-global-weirding; "Global Weirding with Katharine Hayhoe," National Public Radio series, 2020–2021, https://www.npr.org/podcasts/961315153/global-weirding-with-katharine-hayhoe; Angela Herring, "3Qs: What Is 'Global Weirding'?," News@Northeastern, March 20, 2012, https://news.northeastern.edu/2012/03/20/globalweirding/; juliazeh, "Global Weirding: The Basics of Ocean Acidification and Climate Change," *North American Marine Environment Protection Association* (blog), October 25, 2016, https://namepa.net/2016/10/25/2016-10-25-global-weirding-part-1/; John Waldman, "With Temperatures Rising, Here Comes 'Global Weirding,'" Yale Environment 360, March 19, 2009, https://e360.yale.edu/features/with_temperatures_rising_here_comes_global_weirding; Google Books Ngram Viewer, comparison of "climate change," "climate crisis," and "global warming," https://books.google.com/ngrams.
23. Gaston Bachelard, *The Psychoanalysis of Fire*, trans. Alan C. M. Ross (Boston: Beacon Books, 1964). My thanks to Eric Rabkin for this pointer, as well as for so much enlightenment in so many things.
24. "It's Official: July Was Earth's Hottest Month on Record," National Oceanic and Atmospheric Administration, August 13, 2021, https://www.noaa.gov/news/its-official-july-2021-was-earths-hottest-month-on-record; Rebecca Dzombak, "Today's Wildfires Are Taking Us into Uncharted Territory," *Scientific American*, July 20, 2021, https://www.scientificamerican.com/article/todays-wildfires-are-taking-us-into-uncharted-territory/; Andrea Januta and Daniel Trotta, "Hoover Dam Reservoir Hits Record Low, in Sign of Extreme Western U.S. Drought," Reuters, June 10, 2021, https://www.reuters.com/world/us/hoover-dam-reservoir-hits-record-low-sign-extreme-western-us-drought-2021-06-10/; Craig Takeuchi, "Ice Quake Occurs in Alaska near B.C. Border While Earthquake Hits off Oregon Coast," *Georgia Straight* (Vancouver, BC), June 29, 2021, https://www.straight.com

/news/ice-quake-occurs-in-alaska-near-bc-border-while-earthquake-hits-off-oregon-coast; Sara Ryding, Marcel Klaassen, Glenn J. Tattersall, Janet L. Gardner, and Matthew R. E. Symonds, "Shape-Shifting: Changing Animal Morphologies as a Response to Climatic Warming," *Trends in Ecology and Evolution* 36, no. 11 (November 2021): 1036–48, https://www.sciencedirect.com/science/article/abs/pii/S016953472100197X.

25. Svetlana Skarbo, "Kolyma Highway in Yakutia, Also Known as the Road of Bones, Is on Fire and Temporarily Shut," Siberian Times, June 30, 2021, https://siberiantimes.com/other/others/news/kolyma-highway-in-yakutia-also-known-as-the-road-of-bones-is-on-fire-and-temporarily-shut/; Lisa Cox and agencies, "Nordic Countries Endure Heatwave as Lapland Records Hottest Day since 1914," *Guardian*, July 6, 2021, https://www.theguardian.com/environment/2021/jul/06/heatwave-hits-nordic-countries-lapland-temperature-; Lorenzo Tondo and agencies, "Tourists Evacuated from Pescara as Italy Records More than 800 Wildfires," *Guardian*, August 1, 2021, https://www.theguardian.com/world/2021/aug/01/tourists-evacuated-from-pescara-as-italy-records-over-800-wildfires; "Death Toll in Turkey Wildfires Rises to Eight, Coastal Resorts Affected," Reuters, August 2, 2021, https://www.reuters.com/world/middle-east/some-wildfires-rage-turkey-although-most-have-been-contained-2021-08-01/; "Angela Merkel Says Germany Must Do More to Fight Climate Crisis," *Guardian*, July 18, 2021, https://www.theguardian.com/world/2021/jul/18/angela-merkel-to-visit-flood-ravaged-areas-in-germany; Ryan Woo and Stella Qiu, "At Least 25 Dead as Rains Deluge Central China's Henan Province," Reuters, July 21, 2021, https://www.reuters.com/world/china/heavy-rainfall-kills-12-central-chinas-henan-provincial-capital-xinhua-2021-07-20/; "Heatwave Causes Massive Melt of Greenland Ice Sheet," Phys.org, July 31, 2021, https://phys.org/news/2021-07-heatwave-massive-greenland-ice-sheet.html; Robert Monroe, "Coronavirus Response Barely Slows Rising Carbon Dioxide," Scripps Institution of Oceanography, June 7, 2021, https://scripps.ucsd.edu/news/coronavirus-response-barely-slows-rising-carbon-dioxide; "Part of Gulf Stream at Risk as Atlantic Ocean Currents Weaken," Al Jazeera, August 6, 2021, https://www.aljazeera.com/news/2021/8/6/atlantic-ocean-currents-weaken-signalling-big-weather.

26. Isabella Kaminskion, "Norwegian Arctic Oil Drilling Targeted by Campaigners in New Legal Action," DeSmog, June 15, 2021, https://www.desmog.com/2021/06/15/norwegian-arctic-oil-drilling-targeted-by-campaigners-in-new-legal-action/; Oliver Milman, "US Climate Research Outpost"; "World's Largest Plant Capturing Carbon from Air Starts in Iceland," Reuters, September 13, 2021, https://www.reuters.com/business/environment/worlds-largest-plant-capturing-carbon-air-starts-iceland-2021-09-08/; "Indonesian State Coal Miner Plans Solar Power Projects," Reuters, September 1, 2021, https://www.reuters.com/article/bukit-asam-solar/indonesian-state-coal-miner-plans-solar-power-projects-idUSL4N2Q31UY; Phoebe Cooke, "Emissions from Burning Oil 'Not Relevant' to North Sea BP Permit, Says UK Government," DeSmog, September 2, 2021, https://www.desmog.com/2021/09/02/oil-uk-bp-vorlich-north-sea-oil/; Sudarshan Varadhan, "India Asks Utilities to Import Coal amid Short Supply as Demand Spikes," Reuters,

September 1, 2021, https://www.reuters.com/world/india/indias-august-power-output-rises-161-coal-fired-power-by-237-2021-09-01/; Brett Wilkins, "As Big Oil Execs Roam Free, Climate Activist Gets 8 Years in Prison," Common Dreams, July 5, 2021, https://www.commondreams.org/news/2021/07/05/big-oil-execs-roam-free-climate-activist-gets-8-years-prison.

27. Warren Ellis, *Normal* (New York: Macmillan, 2016).

Chapter 1. Uprooting the Campus

Epigraphs. Elaine Newbern, "Rising to the Challenge: Eckerd's Plan to Combat Rising Sea Levels," *Current*, February 27, 2020, http://www.theonlinecurrent.com/news/rising-to-the-challenge-eckerd-s-plan-to-combat-rising-sea-levels/article_c7691d4c-5781-11ea-a65b-a7d50f6ad3cc.html/. Virginia Sapiro, "The Life Course of Higher Education Institutions: When the End Comes," working paper, 2019, https://blogs.bu.edu/vsapiro/files/2019/02/SapiroWhentheEndComes2019-1.pdf.

1. Lee Gardner, "For Colleges, Climate Change Means Making Tough Choices," *Chronicle of Higher Education*, May 3, 2019, https://www.chronicle.com/article/for-colleges-climate-change-means-making-tough-choices/; J. L. Weiss, J. T. Overpeck, and B. Strauss, "Implications of Recent Sea Level Rise Science for Low-Elevation Areas in Coastal Cities of the Conterminous U.S.A," *Climatic Change* 105 (2011): 635–45, https://doi.org/10.1007/s10584-011-0024-x; Ben Myers and Erica Lusk, "Rising Threat as the Climate Changes and Seas Swell, Coastal Colleges Struggle to Prepare," *Chronicle of Higher Education*, December 6, 2017, https://www.chronicle.com/interactives/rising-threat; Li You, "Rising Seas Threaten China's Long, Low, and Crowded Coast," Sixth Tone, December 31, 2020, https://www.sixthtone.com/news/1006652/rising-seas-threaten-chinas-long%2C-low%2C-and-crowded-coast; "CREAT Climate Scenarios Projection Map," Environmental Protection Agency, https://www.arcgis.com/apps/MapSeries/index.html?appid=3805293158d54846a29f750d63c6890e.
2. Adrian D. Wernera, Mark Bakker, Vincent E. A. Post, Alexander Vandenbohede, Chunhui Lu, Behzad Ataie-Ashtiania, Craig T. Simmons, and D. A. Barry, "Seawater Intrusion Processes, Investigation and Management: Recent Advances and Future Challenges," *Advances in Water Resources* 51 (January 2013): 3–26, https://www.sciencedirect.com/science/article/abs/pii/S030917081200053X?via%3Dihub; Stephen F. Jane, Gretchen J. A. Hansen, Benjamin M. Kraemer, Peter R. Leavitt, Joshua L. Mincer, Rebecca L. North, Rachel M. Pilla, et al., "Widespread Deoxygenation of Temperate Lakes," *Nature* 594 (2021): 66–70, https://doi.org/10.1038/s41586-021-03550-y; Mads Nyborg Støstad and Patrick da Silva Sæther, "Something Is Happening to Norway," NRK, December 1, 2019, https://www.nrk.no/chasing-climate-change-1.14859595.
3. A. M. Vicedo-Cabrera, N. Scovronick, F. Sera, D. Royé, R. Schneider, A. Tobias, C. Astrom, et al., "The Burden of Heat-Related Mortality Attributable to Recent Human-Induced Climate Change," *Nature Climate Change* 11 (2021): 492–500, https://doi.org/10.1038/s41558-021-01058-x; Anchal Vohra, "The Middle East Is

Becoming Literally Uninhabitable," *Foreign Policy*, August 24, 2021, https://foreignpolicy.com/2021/08/24/the-middle-east-is-becoming-literally-uninhabitable/; Al Shaw, Abrahm Lustgarten, and Jeremy W. Goldsmith, "New Climate Maps Show a Transformed United States," ProPublica, September 15, 2020, https://projects.propublica.org/climate-migration/; Colin Raymond, Tom Matthews, and Radley M. Horton, "The Emergence of Heat and Humidity Too Severe for Human Tolerance," *Science Advances* 6, no. 19 (May 8, 2020), https://doi.org/10.1126/sciadv.aaw1838; Alexander Dacy, "Olivia Paregol's Family Takes Steps to Sue UMD for Its Response to Adenovirus, Mold," *Diamondback*, May 21, 2019, https://dbknews.com/2019/05/21/umd-olivia-paregol-adenovirus-lawsuit-mold-president-loh/; Simon Lewis, "Canada Is a Warning: More and More of the World Will Soon Be Too Hot for Humans," *Guardian*, June 30, 2021, https://www.theguardian.com/commentisfree/2021/jun/30/canada-temperatures-limits-human-climate-emergency-earth.

4. Haider Ali, Hayley J. Fowler, Geert Lenderink, Elizabeth Lewis, and David Pritchard, "Consistent Large-Scale Response of Hourly Extreme Precipitation to Temperature Variation over Land," *Geophysical Research Letters*, January 12, 2021, https://doi.org/10.1029/2020GL090317; Abdullah Kahraman, Elizabeth J. Kendon, Steven C. Chan, and Hayley J. Fowler, "Quasi-stationary Intense Rainstorms Spread across Europe under Climate Change," *Geophysical Research Letters* 48, no. 13 (June 30, 2021), https://doi.org/10.1029/2020GL092361.
5. Ludovicus P. Van Beek, Inge de Graaf, Edwin Sutanudjaja, Yoshihide Wada, and Marc F. P. Bierkens, "Limits to Global Groundwater Consumption," American Geophysical Union conference presentation, December 15, 2016, https://agu.confex.com/agu/fm16/meetingapp.cgi/Paper/176019.
6. K. A. McKinnon, A. Poppick, and I. R. Simpson, "Hot Extremes Have Become Drier in the United States Southwest," *Nature Climate Change* 11 (2021): 598–604, https://doi.org/10.1038/s41558-021-01076-9; Scripps Institution of Oceanography, *Climate Change and Sea Level Rise: Scenarios for California Vulnerability and Adaptation Assessment*, CEC-500-2012-008 (Sacramento: California Energy Commission, July 2012), https://uc-ciee.org/ciee-old/downloads/Climate%20and%20Sea%20Level%20Rise%20Scenarios%20for%20California.pdf.
7. Nina Lakhani, "Killer Heat: US Racial Injustices Will Worsen as Climate Crisis Escalates," *Guardian*, July 28, 2020, https://www.theguardian.com/us-news/2020/jul/28/us-racial-injustices-will-worsen-climate-crisis-escalates.
8. Climate Central, *Global Weirdness: Severe Storms, Deadly Heat Waves, Relentless Drought, Rising Seas, and the Weather of the Future* (New York: Penguin Random House, 2013), 108–9; James McEldowney, "EU Agricultural Policy and Climate Change," European Parliamentary Research Service, PE 651.922, May 2020, https://www.europarl.europa.eu/RegData/etudes/BRIE/2020/651922/EPRS_BRI(2020)651922_EN.pdf; Richard Seager, Nathan Lis, Jamie Feldman, Mingfang Ting, A. Park Williams, Jennifer Nakamura, Haibo Liu, and Naomi Henderson, "Whither the 100th Meridian? The Once and Future Physical and Human Geography of America's Arid–Humid Divide. Part I: The Story So Far," *Earth*

Interactions 22, no. 5 (March 1, 2018): 1–22, https://doi.org/10.1175/EI-D-17-0011.1; Stefan van der Esch, Ben ten Brink, Elke Stehfest, Michel Bakkenes, Annelies Sewell, Arno Bouwman, Johan Meijer, Henk Westhoek, and Maurits van den Berg, *Exploring Future Changes in Land Use and Land Condition and the Impacts on Food, Water, Climate Change and Biodiversity* (The Hague: PBL Netherlands Environmental Assessment Agency, 2017), https://www.pbl.nl/sites/default/files/downloads/pbl-2017-exploring-future-changes-in-land-use-and-land-condition-2076b.pdf; Conrad Duncan, "Brazil Facing Worst Drought in Nearly 100 Years as Officials Issue Emergency Warning," *Independent*, May 29, 2021, https://www.independent.co.uk/news/world/americas/brazil-drought-water-amazon-rainforest-b1856404.html; Anastasia Moloney, "The Poorest in Guatemala Bear Brunt of Climate Change, Research Says," Reuters, May 3, 2019, https://www.reuters.com/article/us-guatemala-climatechange-poor-idUSKCN1SA024.

9. S. Feng and Q. Fu, "Expansion of Global Drylands under a Warming Climate," *Atmospheric Chemistry and Physics* 13, no. 19 (2013): 10081–94, https://doi.org/10.5194/acp-13-10081-2013; Jianping Huang, Haipeng Yu, Xiaodan Guan, Guoyin Wang, and Ruixia Guo, "Accelerated Dryland Expansion under Climate Change," *Nature Climate Change* 6 (2016): 166–71, https://doi.org/10.1038/nclimate2837; James Temple, "Yes, Climate Change Is Almost Certainly Fueling California's Massive Fires," *MIT Technology Review*, August 20, 2020, https://www.technologyreview.com/2020/08/20/1007478/california-wildfires-climate-change-heatwaves/.
10. Robert Cushman (Pacific Union College president) and Maria Rankin-Brown (academic dean), interview by the author, November 6, 2020.
11. Miyuki Hino and Marshall Burke, "The Effect of Information about Climate Risk on Property Values," *PNAS* 118, no. 17 (April 20, 2021): e2003374118, https://doi.org/10.1073/pnas.2003374118.
12. Andreas Malm, *How to Blow Up a Pipeline: Learning to Fight in a World on Fire* (New York: Verso, 2021); Kim Stanley Robinson, *The Ministry for the Future* (New York: Orbit, 2020).
13. Paul Hawken, *Drawdown: The Most Comprehensive Plan Ever Proposed to Reverse Global Warming* (New York: Penguin, 2017), 164–65; "The Montreal Protocol Evolves to Fight Climate Change," United Nations Industrial Development Organization, https://www.unido.org/our-focus-safeguarding-environment-implementation-multilateral-environmental-agreements-montreal-protocol/montreal-protocol-evolves-fight-climate-change; Eric Dean Wilson, *After Cooling: On Freon, Global Warming, and the Terrible Cost of Comfort* (New York: Simon and Schuster, 2021); Nathan B. Morris, Georgia K. Chaseling, Timothy English, Fabian Gruss, Mohammad Fauzan Bin Maideen, Anthony Capon, and Ollie Jay, "Electric Fan Use for Cooling during Hot Weather: A Biophysical Modelling Study," *Lancet Planetary Health* 5, no. 6 (June 1, 2021): E368–E377, https://doi.org/10.1016/S2542-5196(21)00136-4.
14. Gardner, "For Colleges."
15. Douglas Barnes, "Passive Air-Conditioning and Refrigeration," *Permaculture Reflections*, November 22, 2006, https://www.permaculturereflections.com/passive

-cooling/; Elmira Jamei, Hing Wah Chau, Mehdi Seyedmahmoudian, and Alex Stojcevski, "Review on the Cooling Potential of Green Roofs in Different Climates," *Science of the Total Environment* 791 (October 15, 2021): 148407, https://doi.org/10.1016/j.scitotenv.2021.148407; Hawken, *Drawdown*, 90–91; Stasia Widerynski, Paul Schramm, Kathryn Conlon, Rebecca Noe, Elena Grossman, Michelle Hawkins, Seema Nayak, Matthew Roach, and Asante Shipp Hilts, *The Use of Cooling Centers to Prevent Heat-Related Illness: Summary of Evidence and Strategies for Implementation* (Atlanta: Centers for Disease Control and Prevention National Center for Environmental Health, n.d.), https://www.cdc.gov/climateandhealth/docs/UseOfCoolingCenters.pdf.

16. Daisy Dunne, "Nitrogen Fertiliser Use Could 'Threaten Global Climate Goals,'" *Carbon Brief*, October 7, 2020, https://www.carbonbrief.org/nitrogen-fertiliser-use-could-threaten-global-climate-goals; Sarah B. Schindler, "Banning Lawns," *George Washington Law Review* 394 (2014), http://digitalcommons.mainelaw.maine.edu/faculty-publications/68; "Extinction Rebellion: Protesters Fined over Cambridge College Lawn Damage," BBC, August 19, 2020, https://www.bbc.com/news/uk-england-cambridgeshire-53839744.
17. Jean-Francois Bastin, Yelena Finegold, Claude Garcia, Danilo Mollicone, Marcelo Rezende, Devin Routh, Constantin M. Zohner, and Thomas W. Crowther, "The Global Tree Restoration Potential," *Science* 365, no. 6448 (July 5, 2019): 76–79, https://doi.org/10.1126/science.aax0848; Joseph W. Veldman, Julie C. Aleman, Swanni T. Alvarado, Michael Anderson, Sally Archibald, William Bond, Thomas W. Boutton, et al., "Comment on 'The Global Tree Restoration Potential,'" *Science* 366, no. 6463 (October 18, 2019), https://doi.org/10.1126/science.aay7976; James Temple, "'A Trillion Trees' Is a Great Idea—That Could Become a Dangerous Climate Distraction," *MIT Technology Review*, January 28, 2020, https://www.technologyreview.com/2020/01/28/276052/tree-planting-is-a-great-idea-that-could-become-a-dangerous-climate-distraction/.
18. Neil Grant, Adam Hawkes, Shivika Mittal, and Ajay Gambhir, "Confronting Mitigation Deterrence in Low-Carbon Scenarios," *Environmental Research Letters* 16, no. 6 (June 28, 2021), https://iopscience.iop.org/article/10.1088/1748-9326/ac0749.
19. Libby Solomon, "Parking Spaces to Parklets: PARK(ing) Day Is Back This September," Greater Greater Washington, August 3, 2021, https://ggwash.org/view/82132/parking-spaces-to-parklets-parking-day-is-back-this-september; Henry Grabar, "The Weight," *Slate*, May 21, 2021, https://slate.com/business/2021/05/ford-f150-lightning-electric-weight.html; S. Hardman and G. Tal, "Understanding Discontinuance among California's Electric Vehicle Owners," *Nature Energy* 6 (2021): 538–45, https://doi.org/10.1038/s41560-021-00814-9; Jessica Britton and Jacob Greig, "Let's Share a Ride: Promoting Carpooling on Campus," Spaces4Learning, January 17, 2020, https://spaces4learning.com/articles/2020/01/17/ridesharing-for-campus.aspx?s=FD&oly_enc_id=; Ellen Rosen, "Making Yellow School Buses a Little More Green," *New York Times*, January 22, 2020, updated August 27, 2021, https://www.nytimes.com/2020/01/22/business/energy-environment/electric-school-buses.html.

20. Hawken, *Drawdown*, 196–97.

21. Association for the Advancement of Sustainability in Higher Education, *STARS Technical Manual*, Version 2.2 (Philadelphia: Association for the Advancement of Sustainability in Higher Education, June 2019), https://stars.aashe.org/wp-content /uploads/2019/07/STARS-2.2-Technical-Manual.pdf; "Plusenergiehaus—the Goal Is PlusEnergy," das plusenergiehaus, http://hosting.more-elements.com/MoccaMS /projects/plusenergie/index.php?p=home&pid=8&L=1&host=1; Hawken, *Drawdown*, 84–85, 98.

22. "Bringing Embodied Carbon Upfront: Coordinated Action for the Building and Construction Sector to Tackle Embodied Carbon," World Green Building Council, 2019, https://www.worldgbc.org/sites/default/files/WorldGBC_Bringing_Embodied _Carbon_Upfront.pdf; Oliver Wainwright, "The Case for . . . Never Demolishing Another Building," *Guardian*, January 13, 2020, https://www.theguardian.com /cities/2020/jan/13/the-case-for-never-demolishing-another-building; Mary Meisenzahl, "This Building in Dubai Is the Largest 3D-Printed Structure in the World—and It Took Just 3 Workers and a Printer to Build It," *Business Insider*, December 30, 2019, https://www.businessinsider.com/dubai-largest-3d-printed -building-apis-cor-photos-2019-12.

23. Ayodele Johnson, "How Africa's Largest City Is Staying Afloat," BBC Future Planet, January 21, 2021, https://www.bbc.com/future/article/20210121-lagos -nigeria-how-africas-largest-city-is-staying-afloat; Kate Orff, "Mending the Landscape," in *All We Can Save: Truth, Courage, and Solutions for the Climate Crisis*, ed. Katharine K. Wilkinson and Ayana Elizabeth Johnson, (New York: Penguin, 2020), 180.

24. Danielle Ohl, "Naval Academy Plans to Raise Seawall as Annapolis Sea Level Rise Looms," *Capital Gazette* (Annapolis, MD), December 3, 2018, https://www.capita lgazette.com/maryland/annapolis/ac-cn-academy-seawall-20181203-story.html; EDAW Inc., *Naval Support Facility Annapolis Master Plan Update* (N.p.: EDAW, May 2006), https://www.usna.edu/Academics/_files/documents/MasterPlanUpdate Draft6.26.06_smlr.pdf; Elaine Newbern, "Rising to the Challenge: Eckerd's Plan to Combat Rising Sea Levels," *Current*, February 27, 2020, http://www.theonline current.com/news/rising-to-the-challenge-eckerd-s-plan-to-combat-rising-sea -levels/article_c7691d4c-5781-11ea-a65b-a7d50f6ad3cc.html/.

25. "LEED Rating System," United States Green Building Council, https://www.usgbc .org/leed; "The Sustainability Tracking, Assessment & Rating System," Association for the Advancement of Sustainability in Higher Education, https://stars.aashe.org /; "The Zero Code," Architecture 2030, http://zero-code.org/zero-code/; Passivhaus Institut, https://passivehouse.com/.

26. Chris Soelberg and Julie Rich, "Sustainable Construction Methods Using Ancient BAD GIR (Wind Catcher) Technology," Construction Research Congress 2014, https://ascelibrary.org/doi/10.1061/9780784413517.161; Alexander Walter, "Amazon HQ2 to Feature a Tree-Covered Swirling Glass Tower, 'The Helix,'" Archinect, February 2, 2021, https://archinect.com/news/article/150247951/amazon-hq2-to -feature-a-tree-covered-swirling-glass-tower-the-helix; Jenny Xie, "Vertical

Forests May Help Solve Climate Change and Housing Shortages," Curbed, August 9, 2017, https://archive.curbed.com/2017/8/9/16059384/vertical-forest-italy-climate-change; Noah Smith, "Drawing Pictures of Cities," Substack, July 17, 2021, https://noahpinion.substack.com/p/drawing-pictures-of-cities?s=r.

27. "Major Cuts of Greenhouse Gas Emissions from Livestock within Reach," Food and Agriculture Organization of the United Nations, September 26, 2013, https://www.fao.org/news/story/en/item/197608/icode; Climate and Clean Air Coalition and United Nations Environment Programme, *Global Methane Assessment: Benefits and Costs of Mitigating Methane Emissions* (Nairobi: United Nations Environment Programme, 2021), https://www.ccacoalition.org/en/resources/global-methane-assessment-full-report; Michael A. Clark, Nina G. G. Domingo, Kimberly Colgan, Sumil K. Thakrar, David Tilman, John Lynch, Inês L. Azevedo, and Jason D. Hill, "Global Food System Emissions Could Preclude Achieving the 1.5° and 2°C Climate Change Targets," *Science* 370, no. 6517 (November 6, 2020): 705–8, https://doi.org/10.1126/science.aba7357; "Food in the Anthropocene: The EAT-Lancet Commission on Healthy Diets from Sustainable Food Systems," *Lancet*, January 16, 2019, https://www.thelancet.com/commissions/EAT.
28. Taylor Dafoe, "Pledging Solidarity with Communities Facing 'Climate Extinction,' the Tate Unveils an Ambitious New Sustainability Plan," Artnet News, July 17, 2019, https://news.artnet.com/art-world/tate-climate-emergency-1603438; Damian Carrington, "UK Health Professions Call for Climate Tax on Meat," *Guardian*, November 4, 2020, https://www.theguardian.com/environment/2020/nov/04/uk-health-professions-call-for-climate-tax-on-meat.
29. Emily Atkin, "Against Meatposting: Progressives Who Glorify Meat Consumption Are Doing Free PR for a Highly-Polluting Industry Working Tirelessly to Keep Polluting," Heated, April 8, 2021, https://heated.world/p/stop-meatposting?s=r; "In-Depth Q&A: What Does the Global Shift in Diets Mean for Climate Change?," Carbon Brief, September 15, 2020, https://www.carbonbrief.org/in-depth-qa-what-does-the-global-shift-in-diets-mean-for-climate-change.
30. Kris De Decker, "How Sustainable Is High-Tech Health Care?," Low-Tech Magazine, February 18, 2021, https://solar.lowtechmagazine.com/2021/02/how-sustainable-is-high-tech-health-care.html.
31. Rebecca Tan, "In Liberal Takoma Park, a Bold New Climate Proposal: Banning Fossil Fuels," *Washington Post*, February 21, 2020, https://www.washingtonpost.com/local/md-politics/takoma-park-fossil-fuel-ban/2020/02/20/307f7c44-5341-11ea-929a-64efa7482a77_story.html.
32. For example, see Nico Wunderling, Jonathan F. Donges, Jürgen Kurths, and Ricarda Winkelmann, "Interacting Tipping Elements Increase Risk of Climate Domino Effects under Global Warming," *Earth System Dynamics* 12 (2021): 601–19, https://doi.org/10.5194/esd-12-601-2021.
33. Emma Strubell, Ananya Ganesh, and Andrew McCallum, "Energy and Policy Considerations for Deep Learning in NLP," arXiv:1906.02243 [cs.CL], June 5, 2019, https://doi.org/10.48550/arXiv.1906.02243; Shangrong Jiang, Yuze Li, Quanying Lu, Yongmiao Hong, Dabo Guan, Yu Xiong, and Shouyang Wang, "Policy Assessments

for the Carbon Emission Flows and Sustainability of Bitcoin Blockchain Operation in China," *Nature Communications* 12 (2021): 1938, https://www.nature.com/articles/s41467-021-22256-3; Renee Obringer, Benjamin Rachunok, Debora Maia-Silva, Maryam Arbabzadeh, Roshanak Nateghi, and Kaveh Madani, "The Overlooked Environmental Footprint of Increasing Internet Use," *Resources, Conservation and Recycling* 167 (2021): 105389, https://doi.org/10.1016/j.resconrec.2020.105389.

34. Kris De Decker, "How and Why I Stopped Buying New Laptops" Low-Tech Magazine, December 2020, https://solar.lowtechmagazine.com/2020/12/how-and-why-i-stopped-buying-new-laptops.html.
35. Roy Schwartz, Jesse Dodge, Noah A. Smith, and Oren Etzioni, "Green AI," arXiv:1907.10597v3 [cs.CY], August 13, 2019, https://arxiv.org/pdf/1907.10597.pdf; "Dr. Aras Bozkurt," Higher Education's Big Rethink, https://ldt.georgetown.edu/big-rethink-webinar-series/aras-bozkurt/; "Green Design," Massachusetts Green High Performance Computing Center, https://www.mghpcc.org/green-design/.
36. Kevin Lozano, "Can the Internet Survive Climate Change? How a Warming World Is Sparking Calls for a Greener Web," *New Republic*, December 18, 2019, https://newrepublic.com/article/155993/can-internet-survive-climate-change. For examples of campuses divesting from fossil fuels, see "Macalester College Divests of All Oil, Gas Investments That Includes Enbridge," KSTP, August 25, 2021, updated September 24, 2021, https://kstp.com/kstp-news/local-news/macalester-college-divests-of-all-oil-gas-investments-that-includes-enbridge/; and Matthew Taylor, "Cambridge University to Divest from Fossil Fuels by 2030," *Guardian*, October 1, 2020, https://www.theguardian.com/education/2020/oct/01/cambridge-university-divest-fossil-fuels-2030-climate.
37. Kerrin Thomas, "NSW Bushfires See Charles Sturt University Open Port Macquarie Campus to House Firefighters," ABC, November 27, 2019, https://www.abc.net.au/news/2019-11-28/firefighters-stay-at-charles-sturt-port-macquarie-bushfires/11738678; Elin Johnson, "As Fires Rage, More Campuses Close across California," Inside Higher Ed, October 29, 2019, https://www.insidehighered.com/news/2019/10/29/california-fires-and-power-outages-close-campuses.

Chapter 2. Doing Research in the Anthropocene

Epigraph. Jeff Goodell, *How to Cool the Planet: Geoengineering and the Audacious Quest to Fix Earth's Climate* (New York: Houghton Mifflin Harcourt, 2010), 106.

1. Robert Pollin, "De-growth vs a Green New Deal," *New Left Review* no. 112 (July/August 2018), https://newleftreview.org/issues/ii112/articles/robert-pollin-de-growth-vs-a-green-new-deal.
2. Sayedeh Sara Sayedi, Benjamin W. Abbott, Brett F. Thornton, Jennifer M. Frederick, Jorian E. Vonk, Paul Overduin, Christina Shadel, et al., "Subsea Permafrost Carbon Stocks and Climate Change Sensitivity Estimated by Expert Assessment," *Environmental Research Letters* 15, no. 12 (December 2020), https://iopscience.iop.org/article/10.1088/1748-9326/abcc29; Nico Wunderling, Jonathan F. Donges,

Jürgen Kurths, and Ricarda Winkelmann, "Interacting Tipping Elements Increase Risk of Climate Domino Effects under Global Warming," *Earth System Dynamics* 12 (2021): 601–19, https://doi.org/10.5194/esd-12-601-2021; Guoxiong Zheng, Simon Keith Allen, Anming Bao, Juan Antonio Ballesteros-Cánovas, Matthias Huss, Guoqing Zhang, Junli Li, et al., "Increasing Risk of Glacial Lake Outburst Floods from Future Third Pole Deglaciation," *Nature Climate Change* 11 (2021): 411–17, https://doi.org/10.1038/s41558-021-01028-3; Ariel Miara, Stuart M. Cohen, Jordan Macknick, Charles J. Vörösmarty, Fabio Corsi, Yinong Sun, Vincent C. Tidwell, Robin Newmark, and Balazs M. Fekete, "Climate-Water Adaptation for Future US Electricity Infrastructure," *Environmental Science and Technology* 53, no. 23 (November 20, 2019): 14029–40, https://doi.org/10.1021/acs.est.9b03037; Scott Jasechko and Debra Perrone, "Global Groundwater Wells at Risk of Running Dry," *Science* 372, no. 6540 (April 23, 2021): 418–21, https://doi.org/10.1126/science.abc2755; Kristofer Covey, Fiona Soper, Sunitha Pangala, Angelo Bernardino, Zoe Pagliaro, Luana Basso, Henrique Cassol, et al., "Carbon and Beyond: The Biogeochemistry of Climate in a Rapidly Changing Amazon," *Frontiers in Forests and Global Change* 4 (March 11, 2021), https://doi.org/10.3389/ffgc.2021.618401; Kaj-Ivar van der Wijst, Andries F. Hof, and Detlef P. van Vuuren, "Costs of Avoiding Net Negative Emissions under a Carbon Budget," *Environmental Research Letters* 16, no. 6 (June 10, 2021), https://iopscience.iop.org/article/10.1088/1748-9326/ac03d9.

3. Geert Jan van Oldenborgh, Karin van der Wiel, Sarah Kew, Sjoukje Philip, Friederike Otto, Robert Vautard, Andrew King, et al., "Pathways and Pitfalls in Extreme Event Attribution," *Climatic Change* 166, no. 13 (2021), https://doi.org/10.1007/s10584-021-03071-7; Nikolina Ban, Cécile Caillaud, Erika Coppola, Emanuela Pichelli, Stefan Sobolowski, Marianna Adinolfi, Bodo Ahrens, et al., "The First Multi-model Ensemble of Regional Climate Simulations at Kilometer-Scale Resolution, Part I: Evaluation of Precipitation," *Climate Dynamics* 57 (2021): 275–302, https://doi.org/10.1007/s00382-021-05708-w; Jean-Baptiste Sallée, Violaine Pellichero, Camille Akhoudas, Etienne Pauthenet, Lucie Vignes, Sunke Schmidtko, Alberto Naveira Garabato, Peter Sutherland, and Mikael Kuusela, "Summertime Increases in Upper-Ocean Stratification and Mixed-Layer Depth," *Nature* 591 (2021): 592–98, https://doi.org/10.1038/s41586-021-03303-x; Thomas Slater, Isobel R. Lawrence, Inès N. Otosaka, Andrew Shepherd, Noel Gourmelen, Livia Jakob, Paul Tepes, Lin Gilbert, and Peter Nienow, "Review Article: Earth's Ice Imbalance," *Cryosphere* 15, no. 1 (2021): 233–46, https://doi.org/10.5194/tc-15-233-2021.
4. Nabil Sultan, Andreia Plaza-Faverola, Sunil Vadakkepuliyambatta, Stefan Buenz, and Jochen Knies, "Impact of Tides and Sea-Level on Deep-Sea Arctic Methane Emissions," *Nature Communications* 11 (2020): 5087, https://doi.org/10.1038/s41467-020-18899-3; Kaitlin A. Naughten, Jan De Rydt, Sebastian H. R. Rosier, Adrian Jenkins, Paul R. Holland, and Jeff K. Ridley, "Two-Timescale Response of a Large Antarctic Ice Shelf to Climate Change," *Nature Communications* 12 (2021): 1991, https://doi.org/10.1038/s41467-021-22259-0; P. R. Goode, E. Pallé, A. Shoumko, S. Shoumko, P. Montañes-Rodriguez, S. E. Koonin, "Earth's Albedo 1998–2017 as

Measured from Earthshine," *Geophysical Research Letters*, August 29, 2021, https://doi.org/10.1029/2021GL094888.

5. Kristen Pope, "Melting Ice Opens Doors for Wider Spread of Contaminants, Diseases," Yale Climate Connections, June 4, 2020, https://yaleclimateconnections.org/2020/06/melting-ice-opens-doors-for-wider-spread-of-contaminants-diseases/; "Climate Change Could Cause Sudden Biodiversity Losses Worldwide," UConn Today, April 8, 2020, https://today.uconn.edu/2020/04/climate-change-cause-sudden-biodiversity-losses-worldwide/; John F. Piatt, Julia K. Parrish, Heather M. Renner, Sarah K. Schoen, Timothy T. Jones, Mayumi L. Arimitsu, Kathy J. Kuletz, Barbara Bodenstein, and Marisol García-Reyes, "Extreme Mortality and Reproductive Failure of Common Murres Resulting from the Northeast Pacific Marine Heatwave of 2014–2016," *PLoS One*, January 15, 2020, https://doi.org/10.1371/journal.pone.0226087; David L. Wagner, Eliza M. Grames, Matthew L. Forister, May R. Berenbaum, and David Stopak, "Insect Decline in the Anthropocene: Death by a Thousand Cuts," *PNAS* 118, no. 2 (January 11, 2021): e2023989118, https://doi.org/10.1073/pnas.2023989118.
6. Michael A. Clark, Nina G. G. Domingo, Kimberly Colgan, Sumil K. Thakrar, David Tilman, John Lynch, Inês L. Azevedo, and Jason D. Hill, "Global Food System Emissions Could Preclude Achieving the 1.5° and 2°C Climate Change Targets," *Science* 370, no. 6517 (November 6, 2020): 705–8, https://doi.org/10.1126/science.aba7357; *NOVA*, "Can We Cool the Planet?," premiered October 28, 2020 on PBS, https://www.pbs.org/wgbh/nova/video/can-we-cool-the-planet/; David J. Mildrexler, Logan T. Berner, Beverly E. Law, Richard A. Birdsey, and William R. Moomaw, "Large Trees Dominate Carbon Storage in Forests East of the Cascade Crest in the United States Pacific Northwest," *Frontiers in Forests and Global Change* 3 (November 5, 2020), https://doi.org/10.3389/ffgc.2020.594274; Francesco N. Tubiello, Giulia Conchedda, Nathan Wanner, Sandro Federici, Simone Rossi, and Giacomo Grassi, "Carbon Emissions and Removals from Forests: New Estimates, 1990–2020," *Earth System Science Data* 13, no. 4 (2021): 1681–91, https://doi.org/10.5194/essd-13-1681-2021; Nancy L. Harris, David A. Gibbs, Alessandro Baccini, Richard A. Birdsey, Sytze de Bruin, Mary Farina, Lola Fatoyinbo, et al., "Global Maps of Twenty-First Century Forest Carbon Fluxes," *Nature Climate Change* 11 (2021): 234–40, https://doi.org/10.1038/s41558-020-00976-6.
7. Tina Casey, "Plot Thickens around Floating Offshore Wind Farm Mystery Tour," CleanTechnica, March 1, 2020, https://cleantechnica.com/2020/03/01/plot-thickens-around-floating-offshore-wind-farm-mystery-tour/; Xiangyu Li, Joseph Peoples, Peiyan Yao, and Xiulin Ruan, "Ultrawhite $BaSO_4$ Paints and Films for Remarkable Daytime Subambient Radiative Cooling," *Applied Materials and Interfaces* 13, no. 18 (2021): 21733–39, https://doi.org/10.1021/acsami.1c02368; Nathanael Johnson, "What If We Let the Oceans into Our Cities?," Grist, January 3, 2020, https://grist.org/climate/what-if-we-let-the-oceans-into-our-cities/; Hailey M. Summers, Evan Sproul, and Jason C. Quinn, "The Greenhouse Gas Emissions of Indoor Cannabis Production in the United States," *Nature Sustainability* 4 (2021): 644–50, https://doi.org/10.1038/s41893-021-00691-w; "Ultra-Large Wind Turbine," ARPA-E, https://arpa

-e.energy.gov/technologies/projects/ultra-large-wind-turbine; Annie Sneed, "World's Largest Wind Turbine Would Be Taller than the Empire State Building," *Scientific American*, June 26, 2017, https://www.scientificamerican.com/article/world-rsquo-s-largest-wind-turbine-would-be-taller-than-the-empire-state-building/#; Kais Siala, Afm Kamal Chowdhury, Thanh Duc Dang, and Stefano Galelli, "Solar Energy and Regional Coordination as a Feasible Alternative to Large Hydropower in Southeast Asia," *Nature Communication* 12 (2021): 4159, https://doi.org/10.1038/s41467-021-24437-6.

8. Scott Hardman and Gil Tal, "Understanding Discontinuance among California's Electric Vehicle Owners," *Nature Energy* 6 (2021): 538–45, https://doi.org/10.1038/s41560-021-00814-9; Gwyn Topham, "Fatal Derailment after Heavy Rains Exposes Growing Danger Posed by Extreme Weather," *Guardian*, August 12, 2020, https://www.theguardian.com/world/2020/aug/12/stonehaven-tragedy-shows-rail-industry-must-adapt-to-climate-crisis; C. Brand, J. Anable, and C. Morton, "Lifestyle, Efficiency and Limits: Modelling Transport Energy and Emissions Using a Socio-technical Approach," *Energy Efficiency* 12 (2019): 187–207, https://doi.org/10.1007/s12053-018-9678-9; David Rolnick, Priya L. Donti, Lynn H. Kaack, Kelly Kochanski, Alexandre Lacoste, Kris Sankaran, Andrew Slavin Ross, et al., "Tackling Climate Change with Machine Learning," arXiv:1906.05433v2 [cs.CY], November 5, 2019, https://arxiv.org/pdf/1906.05433.pdf; Jonathan Koomey and Eric Masanet, "Does Not Compute: Avoiding Pitfalls Assessing the Internet's Energy and Carbon Impacts," *Joule* 5, no. 7 (June 24, 2021): 1625–28, https://doi.org/10.1016/j.joule.2021.05.007; Emma Strubell, Ananya Ganesh, and Andrew McCallum, "Considerations for Deep Learning in NLP," arXiv:1906.02243v1 [cs.CL], June 2019, https://doi.org/10.48550/arXiv.1906.02243; Emily M. Bender, Timnit Gebru, Angelina McMillan-Major, and Shmargaret Shmitchell, "On the Dangers of Stochastic Parrots: Can Language Models Be Too Big?," *FAccT '21: Proceedings of the 2021 ACM Conference on Fairness, Accountability, and Transparency* (March 2021): 610–23, https://doi.org/10.1145/3442188.3445922; Renee Obringer, Benjamin Rachunok, Debora Maia-Silva, Maryam Arbabzadeh, Roshanak Nateghic, and Kaveh Madanie, "The Overlooked Environmental Footprint of Increasing Internet Use," *Resources, Conservation and Recycling* 167 (April 2021): 105389, https://doi.org/10.1016/j.resconrec.2020.105389; Roy Schwartz, Jesse Dodge, Noah A. Smith, and Oren Etzioni, "Green AI," *Communications of the ACM* 63, no. 12 (December 2020): 54–63, https://doi.org/10.1145/3381831.
9. Alexander Feher, Martin Hauptvogl, Petra Tangosova, and Lucia Svetlanska, "How to Educate for a Healthy Climate at a University? An Intergenerational Cooperation (a Case Study from Slovakia)," in *Universities and Climate Change: Introducing Climate Change to University Programmes*, ed. Walter Leal Filho (Heidelberg, Germany: Springer, 2010), 144.
10. Steve Keen, "The Cost of Climate Change: A Nobel Economist's Model Dismantled," *Evonomics* (blog), July 14, 2019, https://evonomics.com/steve-keen-nordhaus-climate-change-economics/; Hongbo Duan, Sheng Zhou, Kejun Jiang, and Christoph Bertram, "Assessing China's Efforts to Pursue the 1.5°C Warming

Limit," *Science* 372, no. 6540 (April 23, 2021): 378–85, https://doi.org/10.1126/science.aba8767; Silvia Pianta, Elina Brutschin, Basvan Ruijven, and Valentina Bosettia, "Faster or Slower Decarbonization? Policymaker and Stakeholder Expectations on the Effect of the COVID-19 Pandemic on the Global Energy Transition," *Energy Research and Social Science* 76 (June 2021): 102025, https://doi.org/10.1016/j.erss.2021.102025; Stefano Giglio and Bryan T. Kelly, "Researchers Propose New Method to Hedge against the Risk of Climate Disaster," Yale Insights, October 29, 2019, https://insights.som.yale.edu/insights/researchers-propose-new-method-to-hedge-against-the-risk-of-climate-disaster; Stefano Carattini, Garth Heutel, and Givi Melkadze, *Climate Policy, Financial Frictions, and Transition Risk*, working paper 28525 (Cambridge, MA: National Bureau of Economic Research, March 2021), https://doi.org/10.3386/w28525; James K. Boyce, "Carbon Pricing: Effectiveness and Equity," *Ecological Economics* 150 (August 2018): 52–61, https://doi.org/10.1016/j.ecolecon.2018.03.030.

11. Meng Jiang, Paul Behrens, Tao Wang, Zhipeng Tang, Yadong Yu, Dinjiang Chen, Lin Liu, et al., "Provincial and Sector-Level Material Footprints in China," *PNAS* 116, no. 52 (December 16, 2019): 26484–90, https://doi.org/10.1073/pnas.1903028116; Kate Raworth, *Doughnut Economics: Seven Ways to Think Like a 21st-Century Economist* (New York: Random House, 2017); Jason Hickel, *Less Is More: How Degrowth Will Save the World* (New York: Penguin Random House, 2020).
12. "Independent Expert Panel for the Legal Definition of Ecocide Commentary and Core Text," Stop Ecocide Foundation, June 2021, https://static1.squarespace.com/static/5ca2608ab914493c64ef1f6d/t/60d1e6e604fae2201d03407f/1624368879048/SE+Foundation+Commentary+and+core+text+rev+6.pdf; Ecocide Law, https://ecocidelaw.com/, offers many ecocide law examples and much commentary.
13. Geoff Mann and Joel Wainwright, *Climate Leviathan: A Political Theory of Our Planetary Future* (New York: Verso, 2018). For a good example of the expanded state idea, see Andreas Malm, *Corona, Climate, Chronic Emergency: War Communism in the Twenty-First Century* (New York: Verso, 2020); Olúfẹ́mi O. Táíwò and Beba Cibralic, "The Case for Climate Reparations," Foreign Policy, October 10, 2020, https://foreignpolicy.com/2020/10/10/case-for-climate-reparations-crisis-migration-refugees-inequality/; David Fishman, "China's Advance into the Antarctic," *Lawfare* (blog), October 27, 2019, https://www.lawfareblog.com/chinas-advance-antarctic; Giulia Sciorati, ed., "The Global Race for Antarctica: China vs the Rest of the World?," Istituto per gli Studi di Politica Internazionale, August 23, 2019, https://www.ispionline.it/it/pubblicazione/global-race-antarctica-china-vs-rest-world-23524.
14. Kate Knuth, "Becoming a Climate Citizen," in *All We Can Save: Truth, Courage, and Solutions for the Climate Crisis*, ed. Katharine K. Wilkinson and Ayana Elizabeth Johnson (New York: Penguin, 2020), 129–35; Leah Cardamore Stokes, "A Field Guide for Transformation," in Wilkinson and Johnson, *All We Can Save*, 337–47.
15. Susan Clayton Whitmore-Williams, Christie Manning, Kirra Krygsman, and Meighen Speiser, "Mental Health and Our Changing Climate: Impacts, Implica-

tions, and Guidance," American Psychological Association, March 2017, https://www.apa.org/news/press/releases/2017/03/mental-health-climate.pdf.

16. Gaia Vince, "How Scientists Are Coping with 'Ecological Grief,'" *Guardian*, January 12, 2020, https://www.theguardian.com/science/2020/jan/12/how-scientists-are-coping-with-environmental-grief; "Mental Health and Our Changing Climate: Impacts, Implications, and Guidance," American Psychological Association, March 2017, https://www.apa.org/news/press/releases/2017/03/mental-health-climate.pdf; Emily Apter, "Planetary Dysphoria," *Third Text* 27, no. 1 (January 15, 2013): 131–40, https://doi.org/10.1080/09528822.2013.752197; Glenn Albrecht, Gina-Maree Sartore, Linda Connor, Nick Higginbotham, Sonia Freeman, Brian Kelly, Helen Stain, Anne Tonna, and Georgia Pollard, "Solastalgia: The Distress Caused by Environmental Change," *Australasian Psychiatry* 15, no. S1 (February 1, 2007): S95–S98, https://doi.org/10.1080/10398560701701288; Ashlee Cunsolo, Sherilee L. Harper, Kelton Minor, Katie Hayes, Kimberly G. Williams, and Courtney Howard, "Ecological Grief and Anxiety: The Start of a Healthy Response to Climate Change?," *Lancet: Planetary Health* 4, no. 7 (July 2020), https://www.thelancet.com/action/showPdf?pii=S2542-5196%2820%2930144-3; Ash Sanders, "Under the Weather," in Wilkinson and Johnson, *All We Can Save*, 235, 236; Robert Macfarlane, *Underland: A Deep Time Journey* (New York: W. W. Norton, 2019), 113.
17. Matthew Ballew, Edward Maibach, John Kotcher, Parrish Bergquist, Seth Rosenthal, Jennifer Marlon, and Anthony Leiserowitz, "Which Racial/Ethnic Groups Care Most about Climate Change?," Center for Climate Change Communication, George Mason University, 2020, https://www.climatechangecommunication.org/all/which-racial-ethnic-groups-care-most-about-climate-change/#_edn1; Julius Alexander McGee, Patrick Trent Greiner, and Carl Appleton, "Locked into Emissions: How Mass Incarceration Contributes to Climate Change," *Social Currents*, November 25, 2020, https://doi.org/10.1177/2329496520974006; "Confronting Carbon Inequality: Putting Climate Justice at the Heart of the COVID-19 Recovery," Oxfam, September 21, 2020, https://oxfamilibrary.openrepository.com/bitstream/handle/10546/621052/mb-confronting-carbon-inequality-210920-en.pdf; Matthew Schneider-Mayerson and Leong Kit Ling, "Eco-reproductive Concerns in the Age of Climate Change," *Climatic Change* 163 (2020): 1007–23, https://link.springer.com/article/10.1007/s10584-020-02923-y.
18. Rebecca Klein and Caroline Preston, "Are We Ready? How We Are Teaching—and Not Teaching—Kids about Climate Change," Hechinger Report, May 23, 2020, https://hechingerreport.org/are-we-ready-how-we-are-teaching-and-not-teaching-kids-about-climate-change/; Chantelle Rijs and Frederick Fenter, "The Academic Response to COVID-19," *Frontiers in Public Health*, October 28, 2020, https://doi.org/10.3389/fpubh.2020.621563; C. J. Ryan and Christopher Marsicano, "Examining the Impact of Divestment from Fossil Fuels on University Endowments," *New York University Journal of Law and Business* 17 (2020): 95–152, Roger Williams Univ. Legal Studies Paper No. 195, posted December 18, 2019, last revised February 16, 2021, https://papers.ssrn.com/sol3/papers.cfm?abstract_id=3501231;

Niina Kautto, Alexei Trundle, and Darryn McEvoy, "Climate Adaptation Planning in the Higher Education Sector," *International Journal of Sustainability in Higher Education* 19, no. 7 (2018), https://www.emerald.com/insight/content/doi/10.1108/IJSHE-02-2018-0028/full/html.

19. Adam Trexler and Adeline Johns-Putra, "Climate Change in Literature and Literary Criticism," *WIREs Climate Change* 2, no. 2 (March 2011):185–200, https://doi.org/10.1002/wcc.105; Luke Collins and Brigitte Nerlich, "Examining User Comments for Deliberative Democracy: A Corpus-Driven Analysis of the Climate Change Debate Online," *Environmental Communication* 9, no. 2 (2015): 189–207, https://doi.org/10.1080/17524032.2014.981560.
20. M. C. Fitzpatrick and R. R. Dunn, "Contemporary Climatic Analogs for 540 North American Urban Areas in the Late 21st Century," *Nature Communications* 10, no. 614 (2019), https://doi.org/10.1038/s41467-019-08540-3; David Wallace-Wells, "Global Apathy toward the Fires in Australia Is a Scary Portent for the Future," *New York Magazine*, December 31, 2019, https://nymag.com/intelligencer/2019/12/new-south-wales-fires-in-australia-the-worlds-response.html.
21. "Kathleen Fitzpatrick," Higher Education's Big Rethink, https://ldt.georgetown.edu/big-rethink-webinar-series/kathleen-fitzpatrick/; Kathleen Fitzpatrick, *Generous Thinking: A Radical Approach to Saving the University* (Baltimore: Johns Hopkins University Press, 2021).
22. "Special Issue 2021: A Review of Media Coverage of Climate Change and Global Warming in 2021," Media and Climate Change Observatory, 2021, https://sciencepolicy.colorado.edu/icecaps/research/media_coverage/summaries/special_issue_2021.html; Silvia Pianta and Matthew R. Sisco, "A Hot Topic in Hot Times: How Media Coverage of Climate Change Is Affected by Temperature Abnormalities," *Environmental Research Letters* 15, no. 11 (November 24, 2020): https://iopscience.iop.org/article/10.1088/1748-9326/abb732; "Bolivar Roads Gate System: Coastal Texas Study," ArcGIS Storymaps, July 21, 2021, https://storymaps.arcgis.com/stories/f4e456a3b3ae4c8e8ac9014099b3a52e; Emily Atkin, "Truth Be Told" in Wilkinson and Johnson, *All We Can Save*, 113–20; Christian Parenti, *Tropic of Chaos: Climate Change and the New Geography of Violence* (New York: Bold Type Books, 2011); Kevin Maher, "A Year Later, a Shroud of Uncertainty Still Surrounds UF's COVID-19 Numbers," *Independent Florida Alligator*, April 19, 2021, https://www.alligator.org/article/2021/04/a-year-later-a-shroud-of-uncertainty-still-surrounds-ufs-covid-19-numbers; Wolfgang Blau, "If You're Not a Climate Reporter Yet, You Will Be: COVID-19 Coverage Offers Lessons for Reporting on the Climate Crisis," NiemanLab, July 14, 2021, https://www.niemanlab.org/2021/07/if-youre-not-a-climate-reporter-yet-you-will-be-covid-19-coverage-offers-lessons-for-reporting-on-the-climate-crisis/.
23. "Climate Crisis: The Unsustainable Use of Online Video: Our New Report on the Environmental Impact of ICT," Shift Project, July 11, 2019, updated June 2020, https://theshiftproject.org/en/article/unsustainable-use-online-video/; Arman Shehabi, Ben Walker, and Eric Masanet, "The Energy and Greenhouse-Gas Implications of Internet Video Streaming in the United States," *Environmental*

Research Letters 9, no. 5, May 28, 2014, https://iopscience.iop.org/article/10.1088/1748-9326/9/5/054007; "Factcheck: What Is the Carbon Footprint of Streaming Video on Netflix?," CarbonBrief, February 25, 2020, https://www.carbonbrief.org/factcheck-what-is-the-carbon-footprint-of-streaming-video-on-netflix.

24. Jem Bendell and Katie Carr, "The Love in Deep Adaptation—a Philosophy for the Forum," *Professor Jem Bendell* (blog), March 17, 2019, https://jembendell.com/2019/03/17/the-love-in-deep-adaptation-a-philosophy-for-the-forum/; "Encyclical Letter Laudato Si' of the Holy Father Francis on Care for Our Common Home," Holy See, 2015, https://www.vatican.va/content/francesco/en/encyclicals/documents/papa-francesco_20150524_enciclica-laudato-si.html; Mann and Wainwright, *Climate Leviathan*, 179–80; Ibrahim Ozdemir, "What Does Islam Say about Climate Change and Climate Action?," Al Jazeera, August 12, 2020, https://www.aljazeera.com/opinions/2020/8/12/what-does-islam-say-about-climate-change-and-climate-action/; Chen Xia and Martin Schönfeld, "A Daoist Response to Climate Change," *Journal of Global Ethics* 7, no. 2 (2011): 195–203, https://doi.org/10.1080/17449626.2011.590279; Randolph Haluza-DeLay, "Religion and Climate Change: Varieties in Viewpoints and Practices," *WIREs Climate Change* 5, no. 2 (March/April 2014): 261–79, https://doi.org/10.1002/wcc.268.
25. N. Stern, "Ethics, Equity and the Economics of Climate Change; Paper 1: Science and Philosophy," *Economics and Philosophy* 30, no. 3 (2014): 397–444, https://doi.org/10.1017/S0266267114000297; Ruth Irwin, ed., *Climate Change and Philosophy: Transformational Possibilities* (New York: Continuum, 2010); Olúfẹ́mi O. Táíwò, "Climate Colonialism and Large-Scale Land Acquisitions," *Carnegie Climate Governance Initiative* (blog), September 29, 2019, https://www.c2g2.net/climate-colonialism-and-large-scale-land-acquisitions/; Bernard Feltz, "The Philosophical and Ethical Issues of Climate Change," *UNESCO Courier* (2019), https://en.unesco.org/courier/2019-3/philosophical-and-ethical-issues-climate-change.
26. Eunice Foote, "Circumstances Affecting the Heat of the Sun's Rays," *American Journal of Science and the Arts* (1856): 382–88; Wilkinson and Johnson, *All We Can Save* offers a fine, wide-ranging example of women's studies applied to the topic. A good bibliography comes from Michael Svoboda, "A Reading List on Women and Climate Change," Yale Climate Connections, March 27, 2019, https://yaleclimateconnections.org/2019/03/a-reading-list-on-women-and-climate-change/; Miriam Gay-Antaki and Diana Liverman, "Climate for Women in Climate Science: Women Scientists and the Intergovernmental Panel on Climate Change," *PNAS* 115, no. 9 (February 12, 2018): 2060–65, https://doi.org/10.1073/pnas.1710271115.
27. Jonah M. Kessel and Hiroko Tabuchi, "It's a Vast, Invisible Climate Menace. We Made It Visible," *New York Times*, December 12, 2019, https://www.nytimes.com/interactive/2019/12/12/climate/texas-methane-super-emitters.html; Taylor Dafoe, "Artist Anicka Yi Explains Why COVID-19 Is Terrible for Humanity, but Fundamentally 'Good for the Planet,'" Artnet News, July 8, 2020, https://news.artnet.com/art-world/anicka-yi-interview-covid19-1892869; Kristin Toussaint, "Step inside an Apartment from the Climate Change–Ravaged Future," *Fast Company*,

January 27, 2020, https://www.fastcompany.com/90454568/step-inside-an-apartment-from-the-climate-change-ravaged-future; Hannah Steinkopf-Frank, "Inside the Imaginarium of a Solarpunk Architect," Messynessy Chic, June 10, 2021, https://www.messynessychic.com/2021/06/10/inside-the-imaginarium-of-a-solarpunk-architect/; aliciaescott, "Field Study #007, The Extinction Event," Bureau of Linguistic Reality, September 1, 2015, https://bureauoflinguisticalreality.com/2015/09/01/field-study-007-the-extinction-event/.

28. Factum Foundation, *The Aura in The Age of Digital Materiality—Rethinking Preservation in the Shadow of an Uncertain Future* (Bologna: Silvana Editoriale, 2020), https://www.factum-arte.com/resources/files/ff/publications_PDF/the_aura_in_the_age_of_digital_materiality_factum_foundation_2020_web.pdf; The Art Newspaper, "The Future of Museums, Exhibitions and the Objects They Display," YouTube video, 2020, https://youtu.be/PUO57HHCKoM; "One Place, Many Stories: Madaba, Jordan," Cyrak, https://cyark.org/projects/madaba/Guided-Tour/001_mad-en; Bryan Alexander, *The New Digital Storytelling: Creating Narratives with New Media*, 2nd ed. (Santa Barbara: Praeger, 2017). My thanks to Cliff Lynch and Jo Ellen Parker for provocative conversations on this topic.
29. Mark Jaccard, *The Citizen's Guide to Climate Success: Overcoming Myths That Hinder Progress* (Cambridge: Cambridge University Press, 2020).
30. D. S. Lee, D. W. Fahey, A. Skowron, M. R. Allen, U. Burkhardt, Q. Chen, S. J. Doherty, et al., "The Contribution of Global Aviation to Anthropogenic Climate Forcing for 2000 to 2018," *Atmospheric Environment* 244 (January 1, 2021), https://doi.org/10.1016/j.atmosenv.2020.117834; Cécile A. J. Girardin, Stuart Jenkins, Nathalie Seddon, Myles Allen, Simon L. Lewis, Charlotte E. Wheeler, Bronson W. Griscom, and Yadvinder Malhi, "Nature-Based Solutions Can Help Cool the Planet—If We Act Now," *Nature* 593 (May 12, 2021): 191–94, https://www.nature.com/articles/d41586-021-01241-2; S. Deng, S. Liu, X. Mo, L. Jiang, P. Bauer-Gottwein, "Polar Drift in the 1990s Explained by Terrestrial Water Storage," *Geophysical Research Letters* 48, no. 7 (March 22, 2021), https://doi.org/10.1029/2020GL092114.
31. Alexander Kurganskiy, Simon Creer, Natasha De Vere, Gareth W. Griffith, Nicholas J. Osborne, Benedict W. Wheeler, Rachel N. McInes, et al., "Predicting the Severity of the Grass Pollen Season and the Effect of Climate Change in Northwest Europe," *Science Advances* 7, no. 13 (March 26, 2021), https://doi.org/10.1126/sciadv.abd7658; Kai Liu, Ming Wang, and Tianjun Zhou, "Increasing Costs to Chinese Railway Infrastructure by Extreme Precipitation in a Warmer World," *Transportation Research Part D: Transport and Environment* 93 (April 2021), https://doi.org/10.1016/j.trd.2021.102797; Jillian M. Deinesa, Meagan E. Schipanski, Bill Golden, Samuel C. Zipper, Soheil Nozari, Caitlin Rottler, Bridget Guerrero, and Vaishali Sharda, "Transitions from Irrigated to Dryland Agriculture in the Ogallala Aquifer," *Agricultural Water Management* 233 (April 30, 2020), https://doi.org/10.1016/j.agwat.2020.106061; Eric Larson, Chris Greig, Jesse Jenkins, Erin Mayfield, Andrew Pascale, Chuan Zhang, Joshua Drossman, Robert Williams, Steve Pacala, and Robert Socolow, *Net-Zero America: Potential Pathways, Infrastructure, and Impacts* (Princeton, NJ: Princeton University, December 15, 2020), https://

environmenthalfcentury.princeton.edu/sites/g/files/toruqf331/files/2020-12/Princeton_NZA_Interim_Report_15_Dec_2020_FINAL.pdf.

32. Tamma A. Carleton, Amir Jina, Michael T. Delgado, Michael Greenstone, Trevor Houser, and Solomon M. Hsiang, "Valuing the Global Mortality Consequences of Climate Change Accounting for Adaptation Costs and Benefits," working paper 27599, National Bureau of Economic Research, July 2020, http://www.nber.org/papers/w27599; Jay Fuhrman, Haewon McJeon, Pralit Patel, Scott C. Doney, William M. Shobe, and Andres F. Clarens, "Food–Energy–Water Implications of Negative Emissions Technologies in a +1.5°C Future," *Nature Climate Change* 10, (2020): 920–27, https://doi.org/10.1038/s41558-020-0876-z; Clark et al., "Global Food System Emissions."
33. Astghik Mavisakalyan and Yashar Tarverdi, "Gender and Climate Change: Do Female Parliamentarians Make Difference?," *European Journal of Political Economy* 56, no. C (2019): 151–64, https://ideas.repec.org/a/eee/poleco/v56y2019icp151-164.html; Annika Carlsson Kanyama, Jonas Nässén, and René Benders, "Shifting Expenditure on Food, Holidays, and Furnishings Could Lower Greenhouse Gas Emissions by Almost 40%," *Journal of Industrial Ecology* 25, no. 6 (December 2021): 1602–16, https://doi.org/10.1111/jiec.13176; Schneider-Mayerson and Ling, "Eco-reproductive Concerns"; Fuhrman et al., "Food–Energy–Water Implications"; Strubell, Ganesh, and McCallum, "Considerations for Deep Learning."
34. Gauthier Chapelle, Pablo Servigne, Raphaël Stevens, *Another End of the World Is Possible: Living the Collapse (and Not Merely Surviving It)*, trans. Geoffrey Samuel (Cambridge: Polity Books, 2021); A. López, *Ecomedia Literacy: Integrating Ecology into Media Education* (New York: Routledge, 2021); Pablo Servigne, Raphaël Stevens, Gauthier Chapelle, and Daniel Rodary, "Deep Adaptation Opens Up a Necessary Conversation about the Breakdown of Civilization," Open Democracy, August 3, 2020, https://www.opendemocracy.net/en/oureconomy/deep-adaptation-opens-necessary-conversation-about-breakdown-civilisation/; "Environmental Humanities," Wikipedia, https://en.wikipedia.org/wiki/Environmental_humanities; "Topic: Climate Change," Harvard University Graduate School of Design, https://www.gsd.harvard.edu/topic/climate-change/; Joseph B. Bak-Coleman, Mark Alfano, Wolfram Barfuss, Carl S. Bergstrom, Miguel A. Centeno, Iain D. Couzin, et al., "Stewardship of Global Collective Behavior," *PNAS* 118, no. 27 (June 21, 2021): e2025764118, https://doi.org/10.1073/pnas.2025764118.
35. van der Wijst, Hof, and van Vuuren, "Costs of Avoiding."
36. J. Henrich, S. Heine, and A. Norenzayan, "The Weirdest People in the World?," *Behavioral and Brain Sciences* 33, no. 2–3 (2010): 61–83, https://doi.org/10.1017/S0140525X0999152X; "Openings: Higher Education's Challenge to Change in the Face of the Pandemic, Inequity and Racism," Georgetown Master's Program in Learning, Design and Technology (Higher Education's Big Rethink Initiative), 2021, https://ldt.georgetown.edu/openings-release/; Himani Bhakuni and Seye Abimbola, "Epistemic Injustice in Academic Global Health," *Lancet Global Health* 9, no. 10 (October 1, 2021): E1465–E1470, https://doi.org/10.1016/S2214-109X(21)00301-6.

37. Ken Hiltner, "A Nearly Carbon-Neutral Conference Model Overview/Practical Guide," https://hiltner.english.ucsb.edu/index.php/ncnc-guide/.
38. Sophia Kier-Byfield, "As an Academic, Should I Worry about My Conference Carbon Footprint?," *Guardian*, June 11, 2019, https://www.theguardian.com/education/2019/jun/11/as-academics-should-we-worry-about-our-conference-carbon-footprint; Seth Wynes, Simon D. Donner, Steuart Tannason, and Noni Nabors, "Academic Air Travel Has a Limited Influence on Professional Success," *Journal of Cleaner Production* 226 (2019): 959–67, https://doi.org/10.1016/j.jclepro.2019.04.109.
39. "MLA Convention Statistics," Modern Language Association, https://www.mla.org/Convention/Convention-History/MLA-Convention-Statistics; "SCA Biennial: Distribute 2020," Society for Cultural Anthropology, 2020, https://culanth.org/engagements/events; "Virtual Conference Participation," Comparative and International Education Society, 2020, https://cies2020.org/virtual-conference-participation/; Karl Qualls (@prof4russia), "Benefits of @aseeestudies virtual conference 1. Proper and cheap organic breakfast from my garden (except org eggs from farmers mkt) 2. Sweatpants #ASEEES20," Twitter, November 5, 2020, https://twitter.com/prof4russia/status/1324374083233079298.
40. Lisa Janicke Hinchliffe (@lisalibrarian), "My univ would need to rethink the facility eval criteria. Its pretty hard to show national reputation and potential for international impact w/o air travel. And, it certainly isn't going to stop faculty going to DC to 'meet with ppl,'" Twitter, January 4, 2020, https://twitter.com/lisalibrarian/status/1213599419255402496.
41. Stephen Nash, *Virginia Climate Fever: How Global Warming Will Transform Our Cities, Shorelines, and Forests* (Charlottesville: University of Virginia Press, 2014), chapter 5.
42. J. Rosen, "Sustainability: A Greener Culture," *Nature* (2017): 546, 565–67, https://doi.org/10.1038/nj7659-565a; Elie Dolgin, "How Going Green Can Raise Cash for Your Lab," *Nature* 554 (February 7, 2018): 265–67, https://www.nature.com/articles/d41586-018-01601-5; Kristen Pope, "Scientists Seek to Collect Ice Core Samples before Glaciers and Ice Sheets Melt," Yale Climate Connections, August 13, 2020, https://yaleclimateconnections.org/2020/08/scientists-seek-to-collect-ice-core-samples-before-glaciers-and-ice-sheets-melt/.
43. "Higher Education's Role in Adapting to a Changing Climate," Second Nature, https://secondnature.org/wp-content/uploads/Higher_Education_Role_Adapting_Changing_Climate.pdf; H. Tuba Özkan-Haller, "To Help Address the Climate Problem, Universities Must Rethink the Tenure and Promotion System," Yale Climate Connections, May 28, 2021, https://yaleclimateconnections.org/2021/05/to-help-address-the-climate-problem-universities-must-rethink-the-tenure-and-promotion-system/.
44. Andrew R. C. Marshall, "Icebound: The Climate-Change Secrets of 19th Century Ship's Logs," Reuters, December 11, 2019, https://www.reuters.com/investigates/special-report/climate-change-ice-shiplogs/.

Chapter 3. Teaching to the End of the World

Epigraphs. Alex Steffen, "Discontinuity Is the Job: How Climate Change and the Planetary Crisis Are Changing What Works," Substack, August 9, 2021, https://alexsteffen.substack.com/p/discontinuity-is-the-job?s=r. Holly Schofield, "Halps' Promise," in *Glass and Gardens: Solarpunk Winters*, ed. Sarena Ulibarri (Albuquerque, NM: World Weaver Press, 2020), 35. Robert D. Newman, "The Humanities in the Age of Loneliness," *Los Angeles Review of Books*, August 19, 2019, https://lareviewofbooks.org/article/humanities-age-loneliness/.

1. Walter Leal Filho, "Climate Change at Universities: Results of a World Survey," in *Universities and Climate Change: Introducing Climate Change to University Programmes*, ed. Walter Leal Filho (Heidelberg: Springer, 2010), 2.
2. Angela Lashbrook, "'No Point in Anything Else': Gen Z Members Flock to Climate Careers," *Guardian*, September 6, 2021, https://www.theguardian.com/environment/2021/sep/06/gen-z-climate-change-careers-jobs; *Drilled + Earther Present: The ABCs of Big Oil*, podcast, 2021–ongoing, https://podcasts.apple.com/us/podcast/drilled-earther-present-the-abcs-of-big-oil/id1439735906?i=1000535204962.
3. Climate Dynamics Group, https://climate-dynamics.org/.
4. Wolfgang Blau, "If You're Not a Climate Reporter Yet, You Will Be: COVID-19 Coverage Offers Lessons for Reporting on the Climate Crisis," NiemanLab, July 14, 2021, https://www.niemanlab.org/2021/07/if-youre-not-a-climate-reporter-yet-you-will-be-covid-19-coverage-offers-lessons-for-reporting-on-the-climate-crisis/.
5. Rebecca Klein and Caroline Preston, "Are We Ready? How We Are Teaching—and Not Teaching—Kids about Climate Change," Hechinger Report, May 23, 2020, https://hechingerreport.org/are-we-ready-how-we-are-teaching-and-not-teaching-kids-about-climate-change/.
6. "Environmental Design," University of Colorado Boulder, https://www.colorado.edu/envd/about/majors; Robert Cushman (Pacific Union College president) and Maria Rankin-Brown (academic dean), interview by the author, 2020; "Sustainability Dual Major," University of New Hampshire, https://www.unh.edu/sustainability/student-education-engagement/sustainability-dual-major; "Sustainability Environmental Design," UC Davis, https://www.ucdavis.edu/majors/sustainable-environmental-design/.
7. "Sustainability Studies," School of Public Policy, https://spp.umd.edu/your-education/undergraduate/minors#sus; "Climate Science Minor," Program on Climate Change, https://pcc.uw.edu/education/undergraduate-minor/.
8. "Sustainable Development and Climate Change Management," Haw Hamburg, https://www.haw-hamburg.de/en/ftz-nk/; "Geographic Information and Climate Change, MSc," Swansea University, https://www.swansea.ac.uk/postgraduate/taught/science/geography/msc-geographic-information-and-climate-change/; "Urban Environment, Sustainability and Climate Change," Erasmus University Rotterdam, https://www.eur.nl/en/master/urban-environment-sustainability-and-climate-change; "Master of Climate Change Science and Policy—MCCSP,"

Victoria University of Wellington, https://www.wgtn.ac.nz/explore/postgraduate-programmes/master-of-climate-change-science-and-policy/overview; "Climate Adaptation Governance," University of Groningen, https://www.rug.nl/masters/climate-adaptation-governance/?lang=en; "MA in Climate and Society," Columbia Climate School, https://climatesociety.ei.columbia.edu/; "Climate Change Institute," University of Maine, https://climatechange.umaine.edu/graduate-program/.

9. Robert Thorson, interview by the author, 2021.
10. Rachel Genevieve Chia, "A New Filipino Law Requires All Students to Plant 10 Trees in Order to Graduate," *Business Insider*, May 29, 2019, https://www.insider.com/new-law-states-students-in-the-philippines-must-plant-trees-before-they-can-graduate-2019-5.
11. "Open Loop University," Stanford 2025, http://www.stanford2025.com/open-loop-university.
12. "Teaching Climate Change in the Humanities," https://teachingclimatechange.org/.
13. He Pou a Rangi New Zealand Climate Change Commission, *Ināia Tonu Nei: A Low Emissions Future for Aotearoa* (Wellington: He Pou a Rangi Climate Change Commission, 2021), https://ccc-production-media.s3.ap-southeast-2.amazonaws.com/public/Inaia-tonu-nei-a-low-emissions-future-for-Aotearoa/Inaia-tonu-nei-a-low-emissions-future-for-Aotearoa.pdf; Michelle Gamage, "10 Jobs That Will Help Shape a Zero-Emissions Future," Grist, June 26, 2021, https://grist.org/climate-energy/10-jobs-that-will-help-shape-a-zero-emissions-future/.
14. Brian J. Beatty, *Hybrid-Flexible Course Design: Implementing Student-Directed Hybrid Classes* (Provo, UT: EdTech Books, 2019); "What Is HyFlex Teaching?," Future Trends Forum, YouTube video, June 25, 2020, https://youtu.be/jNHnJnzwXuA; "Stories of HyFlex," Future Trends Forum, YouTube video, September 24, 2020, https://youtu.be/kmr8QZwhKXM.
15. Robin Wall Kimmerer, "Sitting in a Circle," in *Braiding Sweetgrass: Indigenous Wisdom, Scientific Knowledge, and the Teachings of Plants* (Minneapolis: Milkweed Editions, 2013), 223–40.
16. Christopher Schaberg and Ned Randolph, "Teaching through the Doom," Inside Higher Ed, January 30, 2020, https://www.insidehighered.com/views/2020/01/30/teaching-about-environment-precarious-surroundings-opinion.
17. Amy Edwards, "Mt Resilience Uses AR to Help You Prepare for Disaster," *CSIROscope* (blog), October 28, 2020, https://blog.csiro.au/mt-resilience-augmented-reality/.
18. Kelly Hydrick, "A Bee Story," StoryCenter, YouTube video, August 2, 2021, https://youtu.be/GonLGoktSTE; "A Story Map on Climate Migrants," TROP ICSU, https://climatescienceteaching.org/tool/story-map-climate-migrants; Bryan Alexander, *The New Digital Storytelling: Creating Narratives with New Media*, 2nd ed. (Santa Barbara: Praeger, 2017).
19. One foundational text is James Paul Gee, *What Video Games Have to Teach Us about Learning and Literacy* (New York: Palgrave Macmillan, 2003). See also Javier Beneyas, Inmaculada Alonso, David Alba, and Luis Pertierra, "The Impact of

Universities on the Climate Change Process," in Leal Filho, *Universities and Climate Change*, 60.

20. "The C-ROADS Climate Change Policy Simulator," Climate Interactive, https://www.climateinteractive.org/c-roads/; "World Climate Simulation," Climate Interactive, https://www.climateinteractive.org/world-climate-simulation/.
21. Stephanie Pfirman and Joey J. Lee, "EcoChains: Arctic Futures," Arizona State University, 2021, https://etx-nec.s3-us-west-2.amazonaws.com/sandbox/eco-chains-test-build/index.html; T. O'Garra, D. Reckien, S. Pfirman, E. Bachrach Simon, G. A. Bachman, J. Brunacini, and J. J. Lee, "Impact of Gameplay vs. Reading on Mental Models of Social-Ecological Systems: A Fuzzy Cognitive Mapping Approach," *Ecology and Society* 26, no. 2 (2021): 25, https://doi.org/10.5751/ES-12425-260225; S. Pfirman, T. O'Garra, E. Bachrach Simon, J. Brunacini, D. Reckien, J. J. Lee, and E. Lukasiewicz, "'Stickier' Learning through Gameplay: An Effective Approach to Climate Change Education," *Journal of Geoscience Education* 69, no. 2 (2021): 192–206, https://doi.org/10.1080/10899995.2020.1858266.
22. "Terra Nil," Devolver, 2021, https://www.devolverdigital.com/games/terra-nil.
23. Emily Byrnes, "Interview with the Mind behind *In Other Waters*: Gareth Damian Martin," Gamespace, March 3, 2020, https://www.gamespace.com/featured/interview-with-the-mind-behind-in-other-waters-gareth-damian-martin/.
24. "Framework for Information Literacy for Higher Education," Association of College and Research Libraries, 2015, https://www.ala.org/acrl/standards/ilframework; "Digital Literacy in Higher Education, Part II: An NMC Horizon Project Strategic Brief," New Media Consortium, August 22, 2017, https://library.educause.edu/resources/2017/8/digital-literacy-in-higher-education-part-ii-an-nmc-horizon-project-strategic-brief; Alex Grech, ed., *Media, Technology and Education in a Post-Truth Society: From Fake News, Datafication and Mass Surveillance to the Death of Trust* (West Yorkshire, UK: Emerald, 2021); "Climate Migrants," ArcGIS StoryMaps, https://storymaps.arcgis.com/collections/af3858d32f84488f92dfaeef068fff52.
25. Ash Sanders, "Under the Weather," in *All We Can Save: Truth, Courage, and Solutions for the Climate Crisis*, ed. Katharine K. Wilkinson and Ayana Johnson (New York: Penguin, 2020), 244–45; Elizabeth Marks, Caroline Hickman, Panu Pihkala, Susan Clayton, Eric R. Lewandowski, Elouise E. Mayall, Britt Wray, Catriona Mellor, and Lise van Susteren, "Young People's Voices on Climate Anxiety, Government Betrayal and Moral Injury: A Global Phenomenon," *Lancet* preprint, September 7, 2021, https://papers.ssrn.com/sol3/papers.cfm?abstract_id=3918955.
26. Schaberg and Randolph, "Teaching through the Doom."
27. M. T. Demneh and Z. H. Darani, "From Remembering to Futuring: Preparing Children for Anthropocene," *Journal of Environmental Studies and Sciences* 10 (2020): 369–79, https://doi.org/10.1007/s13412-020-00634-5.
28. Robert Thorson, interview, 2021; Roland Hergert, Volker Barth, and Thomas Klenke, "Interdisciplinary and Interfaculty Approaches in Higher Education Capable of Permeating the Complexity of Climate Change," in Leal Filho, *Universities and Climate Change*, 107–15.

29. Isaijah Johnson, "'Solarpunk' and the Pedagogical Value of Utopia," *Journal of Sustainability Education*, May 1, 2020, http://www.susted.com/wordpress/content/solarpunk-the-pedagogical-value-of-utopia_2020_05/.
30. Erica Pandey, "January 15, 2020," Axios Future, January 15, 2020, https://www.axios.com/newsletters/axios-future-6cc055e9-7d0a-4874-a790-f8316337ee84.html.
31. Doug Reilly, "Our Ministry for the Future Reading: Part 2," comment on Bryan Alexander, December 18, 2020, https://bryanalexander.org/book-club/our-ministry-for-the-future-reading-part-2/#comment-168034.
32. "Harvard-Yale Game Ends in Near-Darkness after Climate Change Protest," ESPN, November 23, 2019, https://www.espn.com/college-football/story/_/id/28147080/harvard-yale-game-ends-darkness-climate-change-protest.
33. He Pou a Rangi, *Ināia tonu nei.*
34. After I wrote that sentence, I turned to the Duolingo app on my Android phone for my daily language practice. The app informed me that many people are studying Swedish in Sweden, and that they are mostly refugees.
35. Brian Fagan, *The Attacking Ocean: The Past, Present, and Future of Rising Sea Levels* (New York: Bloomsbury, 2013), 187; Colette Pichon Battle, "An Offering from the Bayou" in Wilkinson and Johnson, *All We Can Save*, 331; Sarah Stillman, "Like the Monarch," in Wilkinson and Johnson, *All We Can Save*, 351; Jake Johnson, "'Epic Failure of Humanity': Global Displaced Population Hits All-Time High," Common Dreams, June 18, 2021, https://www.commondreams.org/news/2021/06/18/epic-failure-humanity-global-displaced-population-hits-all-time-high.
36. Adaptation Learning Network, https://adaptationlearningnetwork.com/; "The 2041 Project," SUNY Oneonta, https://suny.oneonta.edu/science-discovery-center/2041-project.

Chapter 4. The Transformation of Town and Gown

Epigraphs. Naomi Klein, *This Changes Everything: Capitalism vs. the Climate* (New York: Simon and Schuster, 2015), 342. Jack Tchen, Erica Kohl-Arenas, and Eric Hartman, "Decolonial Desires: Is a Third University Possible? A Webinar Conversation with K. Wayne Yang," *Ethnic Studies Review* 43, no. 3 (2020): 57–72, https://doi.org/10.1525/esr.2020.43.3.57.

1. Oliver Milman, "Washington State County Is First in US to Ban New Fossil Fuel Infrastructure," *Guardian*, July 28, 2021, https://www.theguardian.com/us-news/2021/jul/28/washington-state-whatcom-county-ban-fossil-fuel-infrastructure; Ethan Goffman, "Montgomery County's Climate Plan Has Ambitious Goals, but Activists Worry It Lacks Teeth," Greater Greater Washington, August 2, 2021, https://ggwash.org/view/82109/montgomery-county-climate-plan-has-the-ambition-the-crisis-demands-but-activists-worry-it-lacks-teeth; Sarah B. Schindler, "Banning Lawns," *George Washington Law Review* 394 (2014), http://digitalcommons.mainelaw.maine.edu/faculty-publications/68; Paul Hawken, *Drawdown: The Most Comprehensive Plan Ever Proposed to Reverse Global Warming* (New York: Penguin, 2017), 5.
2. Mike De Socio, "The US City That Has Raised $100M to Climate-Proof Its Buildings," *Guardian*, August 19, 2021, https://www.theguardian.com/environment

/2021/aug/19/ithaca-new-york-raised-100m-climate-proof-buildings; Niall Patrick Walsh, "Boston Publishes Radical SCAPE Plans to Combat Climate Change," Archdaily, October 24, 2018, https://www.archdaily.com/904610/boston-publishes-radical-scape-plans-to-combat-climate-change.

3. "Miami-Dade Back Bay Coastal Storm Risk Management Draft Integrated Feasibility Report and Programmatic Environmental Impact Statement," US Army Corps of Engineers Norfolk District, May 29, 2020, https://usace.contentdm.oclc.org/utils/getfile/collection/p16021coll7/id/14453.
4. "Methane Research Series: 16 Studies," Environmental Defense Fund, 2018, https://www.edf.org/climate/methane-research-series-16-studies; Scot M. Miller, Anna M. Michalak, Robert G. Detmers, Otto P. Hasekamp, Lori M. P. Bruhwiler, and Stefan Schwietzke, "China's Coal Mine Methane Regulations Have Not Curbed Growing Emissions," *Nature Communications* 10 (2019): 303, https://doi.org/10.1038/s41467-018-07891-7.
5. Rhoda Wanyenze and Barnabas Nawangwe, "Leadership and Partnerships for a Sustainable Future: Experiences from Makerere University School of Public Health, in Uganda," *IAU Horizons* 26, no. 2 (December 2021), https://iau-aiu.net/IMG/pdf/iau_horizons_vol_26_2.pdf.
6. The Ray, https://theray.org; Hawken, *Drawdown*, 196–97.
7. Kate Marsh, Neely McKee, and Maris Welch, "Opposition to Renewable Energy Facilities in the United States," Sabin Center for Climate Change Law, Columbia Law School, February 2021, https://climate.law.columbia.edu/sites/default/files/content/RELDI%20report%20MBG%202.26.21%20HWA.pdf.
8. "California Teachers Could Be Required to Teach about Climate Change," CBS Sacramento, January 14, 2020, https://sacramento.cbslocal.com/2020/01/14/california-teachers-required-teach-climate-change/.
9. Rick Seltzer, "What Other Colleges Can Learn from UC Berkeley's Fraught Town-Gown Relations," Higher Ed Dive, April 14, 2022, https://www.highereddive.com/news/what-other-colleges-can-learn-from-uc-berkeleys-fraught-town-gown-relation/621629/; "A Look at Town-Gown Tensions at UC Berkeley," The Takeaway, March 29, 2022, https://www.wnycstudios.org/podcasts/takeaway/segments/look-town-gown-tensions-uc-berkeley; Tamma A. Carleton, Amir Jina, Michael T. Delgado, Michael Greenstone, Trevor Houser, and Solomon M. Hsiang, "Valuing the Global Mortality Consequences of Climate Change Accounting for Adaptation Costs and Benefits," working paper 27599, National Bureau of Economic Research, July 2020, http://www.nber.org/papers/w27599.
10. Emma Whitford, "A California Community College Faces Wildfire," Inside Higher Ed, August 31, 2020, https://www.insidehighered.com/news/2020/08/31/cabrillo-college-grapples-nearby-wildfire-employees-lose-homes.
11. "The Economics of Climate Change: Climate Change Poses the Biggest Long-Term Risk to the Global Economy. No Action Is Not an Option," Swiss Re, April 22, 2021, https://www.swissre.com/institute/research/topics-and-risk-dialogues/climate-and-natural-catastrophe-risk/expertise-publication-economics-of-climate-change.html.

12. Ali Allawi and Fatih Birol, "Without Help for Oil-Producing Countries, Net Zero by 2050 Is a Distant Dream," *Guardian*, September 1, 2021, https://www.theguardian.com/commentisfree/2021/sep/01/oil-producing-countries-net-zero-2050-iraq.
13. Adele Peters, "This Innovative Tax Plan Is Designed to Help Cities Pay for Climate Action," *Fast Company*, October 21, 2020, https://www.fastcompany.com/90566250/this-innovative-tax-plan-is-designed-to-help-cities-pay-for-climate-action.
14. Isabelle Anguelovski, James J. T. Connolly, Hamil Pearsall, Galia Shokry, Melissa Checker, Juliana Maantay, Kenneth Gould, Tammy Lewis, Andrew Maroko, and J. Timmons Roberts, "Why Green 'Climate Gentrification' Threatens Poor and Vulnerable Populations," *PNAS* 116, no. 52 (December 26, 2019): 26139–43, https://doi.org/10.1073/pnas.1920490117.
15. "School Enrollment, Tertiary, Female (% Gross)," World Bank, data as of September 2021, https://data.worldbank.org/indicator/SE.TER.ENRR.FE?end=2020&start=1970&view=chart&year=1989; Somayeh Parvazian, Judith Gill, and Belinda Chiera, "Higher Education, Women, and Sociocultural Change: A Closer Look at the Statistics," *SAGE Open*, May 16, 2017, https://doi.org/10.1177/2158244017700230; Hawken, *Drawdown*, 76–77, 81–82.
16. Jeanne Herb and Lisa Auermullera, "Final Project Report: A Seat at the Table: Integrating the Needs and Challenges of Underrepresented and Socially Vulnerable Populations into Coastal Hazards Planning in New Jersey," Rutgers New Jersey Climate Change Resource Center, May 31, 2020, https://njclimateresourcecenter.rutgers.edu/wp-content/uploads/2020/06/A-Seat-at-the-Table-Final-Report-5-31-20.pdf; "Climate Preparedness," Partnership for Community Development, https://hamiltonpcd.org/climate-preparedness-working-group/.
17. Neil Leary, interview by the author, 2021.
18. "Energy Fellows Program," University of Rhode Island Cooperative Extension, https://web.uri.edu/coopext/efp/.
19. Stephen Nash, *Virginia Climate Fever: How Global Warming Will Transform Our Cities, Shorelines, and Forests* (Charlottesville: University of Virginia Press, 2014), 106.
20. Varshini Prakash, "We Are Sunrise" in *All We Can Save: Truth, Courage, and Solutions for the Climate Crisis*, ed. Katharine K. Wilkinson and Ayana Elizabeth Johnson (New York: Penguin, 2020), 190; Emma Whitford, "Fossil Fuel Fight Escalates to State Attorney General," Inside Higher Ed, December 16, 2020, https://www.insidehighered.com/news/2020/12/16/boston-college-students-alumni-take-fossil-fuel-investment-battle-state-attorney.
21. "The Land Relationships Super Collective," Land Relationships Super Collective, http://www.landrelationships.com/introduction.

Chapter 5. Academia in the World

Epigraph. Alexis C. Madrigal, "'The Future Is about Old People, in Big Cities, Afraid of the Sky' and Four Other Intriguing Things: When Otters Attack, the Objects of Brain Interfaces, WWI Diaries, and a Digital Model of a Piano," *Atlantic*, March 17, 2014,

https://www.theatlantic.com/technology/archive/2014/03/the-future-is-about-old
-people-in-big-cities-afraid-of-the-sky/284459/.

1. Aims C. McGuiness, “The States and Higher Education,” in *American Higher Education in the Twenty-First Century: Social, Political, and Economic Challenges*, 4th ed., ed. Michael N. Bastedo, Philip G. Altbach, and Patricia J. Gumport (Baltimore: Johns Hopkins University Press, 2016), 240; “New Vehicle Rebate,” Gouvernement du Québec, 2022, https://vehiculeselectriques.gouv.qc.ca/english
/rabais/ve-neuf/programme-rabais-vehicule-neuf.asp. For one example of states, public universities, and COVID, see Elizabeth Redden, “Tying Colleges’ Hands: Arizona Executive Order Barring Public Colleges from Mandating COVID-19 Vaccines, Testing and Mask Wearing Is Expected to be Codified into Law,” Inside Higher Ed, June 30, 2021, https://www.insidehighered.com/news/2021/06/30/arizona
-public-colleges-cant-require-covid-vaccines-or-masks-or-testing-unvaccinated; Sam Rutherford, “It’s about to Get Harder to Buy a Gaming PC,” Gizmodo, July 27, 2021, https://gizmodo.com/its-about-to-get-harder-to-buy-a-gaming-pc-1847372953.
2. Aman Azhar, “Why Is Texas Allocating Funds for Reducing Air Emissions to Widening Highways?,” Inside Climate News, August 23, 2021, https://insideclimate
news.org/news/23082021/texas-air-pollution-highways/.
3. Emily Beament, “New £15.9 Million Grant Scheme Opens for Landowners to Create Woodlands,” *Evening Standard* (London), June 9, 2021, https://www
.standard.co.uk/news/uk/government-england-parliament-environment-secretary
-george-eustice-b939531.html; Mike Bedigan, “More Than £860M to Be Spent on Flood Protections as Heavy Rain Batters the UK,” *Belfast Telegraph*, July 28 2021, https://www.belfasttelegraph.co.uk/news/uk/more-than-860m-to-be-spent-on
-flood-protections-as-heavy-rain-batters-the-uk-40699906.html; Grace Sihombing and Arys Aditya, “Indonesia May Tax the Rich, Top Polluters for More Revenue,” Bloomberg, May 25, 2021, https://www.bloomberg.com/news/articles
/2021-05-25/indonesia-may-tax-the-rich-top-polluters-for-more-state-revenue.
4. Thomas C. Jorling, “Addressing Climate Change the Smart Way,” New England Board of Higher Education, February 10, 2021, https://nebhe.org/journal/addressing
-climate-change-the-smart-way/.
5. Joel Hills, “Government to Help Fund Four Carbon Capture and Storage Hubs in the UK,” ITV, November 16, 2020, https://www.itv.com/news/2020-11-16
/government-to-help-fund-four-carbon-capture-and-storage-hubs-in-the-uk; “Environment Secretary to Set Out Plans to Restore Nature and Build Back Greener from the Pandemic,” press release, Department for Environment, Food and Rural Affairs, Forestry Commission, and Natural England, May 18, 2021, https://www.gov.uk/government/news/environment-secretary-to-set-out-plans
-to-restore-nature-and-build-back-greener-from-the-pandemic; “Wind for Schools Dashboard,” Wind for Schools, https://openei.org/wiki/Wind_for_Schools
_Portal; Kim Willsher, “France to Ban Some Domestic Flights Where Train Available,” *Guardian*, April 12, 2021, https://www.theguardian.com/business/2021
/apr/12/france-ban-some-domestic-flights-train-available-macron-climate
-convention-mps.

6. Matthew Goldstein and Peter Eavis, "The S.E.C. Moves Closer to Enacting a Sweeping Climate Disclosure Rule," *New York Times*, March 21, 2022, https://www.nytimes.com/2022/03/21/business/sec-climate-disclosure-rule.html.
7. "Xi Jinping Ecological Civilization Thought Research Center Was Established," Xinhuanet, Google Translate, July 7, 2021, http://www.xinhuanet.com/politics/2021-07/07/c_1127632693.htm.
8. Isabelle Gerretsen, "South Korea Proposes Cutting Emissions 40% by 2030," Climate Change News, June 16, 2021, https://www.climatechangenews.com/2021/06/16/south-korea-proposes-cutting-emissions-40-2030/.
9. "China to Limit Carbon Dioxide Emissions of Public Institutions to 400 Mln Tonnes by 2025," Xinhua, April 6, 2021, http://www.xinhuanet.com/english/2021-06/04/c_139988753.htm; "China State Planner Vows Crackdown on Projects with High Energy Use," Reuters, August 17, 2021, https://www.reuters.com/business/sustainable-business/china-state-planner-vows-crackdown-projects-with-high-energy-use-2021-08-17/.
10. Tik Root, "Biden Administration Announces New Energy Star Standards, Plans for Emissions Targets for Federal Buildings," *Washington Post*, May 17, 2021, https://www.washingtonpost.com/climate-solutions/2021/05/17/biden-energy-efficiency/; Andrew Charlton-Perez, "Guest Post: The Climate-Change Gaps in the UK School Curriculum," Carbon Brief, September 15, 2021, https://www.carbonbrief.org/guest-post-the-climate-change-gaps-in-the-uk-school-curriculum; Francie Diep, "Texas A&M University Abruptly Shuttered a Climate Lab, Citing Security Risk," *Chronicle of Higher Education*, April 8, 2022, https://www.chronicle.com/article/texas-a-m-university-abruptly-shuttered-a-climate-lab-citing-security-risk.
11. "German Cabinet Agrees More Ambitious CO2 Cuts before September Election," Reuters, May 12, 2021, https://www.reuters.com/world/europe/german-cabinet-passes-plans-more-ambitious-co2-cuts-govt-source-2021-05-12/; John Schwartz, "In 'Strongest' Climate Ruling Yet, Dutch Court Orders Leaders to Take Action," *New York Times*, December 20, 2019, https://www.nytimes.com/2019/12/20/climate/netherlands-climate-lawsuit.html.
12. Rachel Genevieve Chia, "A New Filipino Law Requires All Students to Plant 10 Trees in Order to Graduate," *Business Insider*, May 29, 2019, https://www.insider.com/new-law-states-students-in-the-philippines-must-plant-trees-before-they-can-graduate-2019-5.
13. Vikram Dodd and Jamie Grierson, "Terrorism Police List Extinction Rebellion as Extremist Ideology," *Guardian*, January 10, 2020, https://www.theguardian.com/uk-news/2020/jan/10/xr-extinction-rebellion-listed-extremist-ideology-police-prevent-scheme-guidance.
14. "Police End Extinction Rebellion Occupation of Norway's Oil Ministry," Reuters, August 23, 2021, https://www.reuters.com/business/environment/extinction-rebellion-occupy-norways-oil-ministry-part-10-day-protest-2021-08-23/; "Global Newsletter #55. Power to the People! Rebellion Returns! Friday, September 10, 2021 by Extinction Rebellion," Extinction Rebellion, September 10, 2021, https://rebellion.global/blog/2021/09/10/global-newsletter-55/.

15. Margaret O. Wilder, Robert G. Varady, Andrea K. Gerlak, Stephen P. Mumme, Karl W. Flessa, Adriana A. Zuniga-Teran, Christopher A. Scott, Nicolás Pineda Pablos, and Sharon B. Megdal, "Hydrodiplomacy and Adaptive Governance at the U.S.-Mexico Border: 75 Years of Tradition and Innovation in Transboundary Water Management," *Environmental Science and Policy* 112 (October 2020): 189–202, https://www.sciencedirect.com/science/article/pii/S1462901120301052?via%3Dihub.
16. Alliance of Small Island States, https://www.aosis.org/.
17. "G7 Agree on 'Historic Steps' to Make Climate Reporting Mandatory," Edie, June 6, 2021, https://www.edie.net/g7-agree-on-historic-steps-to-make-climate-reporting-mandatory/; Karl Mathiesen and David M. Herszenhorn, "G7 Wrestles with Its Climate Limitations," Politico, June 11, 2021, https://www.politico.eu/article/g7-wrestles-with-its-climate-limitations-g20-carbon-emissions/.
18. Kate Abnett, "EU Drafts Plan to Grow 'Carbon Sinks' in Climate Change Fight," Reuters, July 6, 2021, https://www.reuters.com/world/europe/eu-drafts-plan-grow-carbon-sinks-climate-change-fight-2021-07-06/; "Europe Has an Ambitious New Climate Plan That Imagines a Dramatic Cut in Emissions," National Public Radio, July 14, 2021, https://www.npr.org/2021/07/14/1015984262/europe-has-an-ambitious-new-climate-plan-that-imagines-a-dramatic-cut-in-emissio; Ewa Krukowska and Alberto Nardelli, "Airlines to Be Charged More for Polluting in EU Green Push," Bloomberg, July 5, 2021, https://www.bloomberg.com/news/articles/2021-07-05/airlines-to-be-charged-more-for-polluting-in-eu-green-push; "EU Commits €4 Billion More to Climate Vulnerables, Calls on the US to Step Up," Climate Home News, September 9, 2021, https://www.climatechangenews.com/2021/09/15/eu-commits-e4-billion-climate-vulnerables-calls-us-step/.
19. "Results," Great Green Wall, https://www.greatgreenwall.org/results.
20. Abiodun Williams, "The United Nations' Leading Role in Tackling the Climate Emergency," *Guardian*, July 6, 2021, https://www.theguardian.com/environment/2021/jul/06/the-united-nations-leading-role-in-tackling-the-climate-emergency; Catherine Philp, "G7 Summit: Broken Promises of Rich Nations Casts Shadow over Climate Deal, Says UN Chief António Guterres," *The Times*, June 10, 2021, https://www.thetimes.co.uk/article/g7-summit-broken-promises-of-rich-nations-casts-shadow-over-climate-deal-says un chicf-antonio-guterres-r37zdd83f; "Climate Change and Disaster Displacement," United Nations High Commissioner for Refugees, https://www.unhcr.org/en-us/climate-change-and-disasters.html; "Displaced on the Frontlines of the Climate Emergency," ArcGIS StoryMap, https://storymaps.arcgis.com/stories/065d18218b654c798ae9f360a626d903; Lisa Schlein, "'Qualification Passport' Enables Refugees to Study, Work in Countries of Exile," Voice of America News, December 22, 2019, https://www.voanews.com/a/middle-east_qualification-passport-enables-refugees-study-work-countries-exile/6181491.html.
21. Kais Siala, Afm Kamal Chowdhury, Thanh Duc Dang, and Stefano Galelli, "Solar Energy and Regional Coordination as a Feasible Alternative to Large Hydropower in Southeast Asia," *Nature Communication* 12 (2021): 4159, https://doi.org/10.1038

/s41467-021-24437-6"; "Consolidated Version of the Energy Charter Treaty and Related Documents," Energy Charter, https://www.energychartertreaty.org/treaty/energy-charter-treaty/; Elizabeth Roche, "India's Green Projects to Get UK Funds," Mint, September 2, 2021, https://www.livemint.com/news/world/uk-india-announce-new-steps-to-tackle-climate-change-11630590540477.html; Mayank Bhardwaj, "India Says to Exceed Emission Cut Targets, Further Reduction Hinges on Climate Fund," Reuters, August 20, 2021, https://www.reuters.com/world/india/india-says-exceed-emission-cut-targets-further-reduction-hinges-climate-fund-2021-08-20/; Sam Meredith, "'We Need to Stop': Inside the World's First Diplomatic Alliance to Keep Oil and Gas in the Ground," CNBC, September 20, 2020, https://www.cnbc.com/2021/09/20/oil-and-gas-inside-the-diplomatic-alliance-to-keep-fossil-fuels-in-the-ground.html; Zack Colman, "China's Xi Pledges to End Funding for Overseas Coal Power Plants," Politico, September 21, 2021, https://www.politico.com/news/2021/09/21/chinas-xi-pledges-to-end-funding-for-overseas-coal-power-plants-513493.

22. Nick O'Malley, "Should We Pay a Carbon Tax to Our Own Government or to Someone Else's?," *Sydney Morning Herald*, June 16, 2021, https://www.smh.com.au/environment/climate-change/should-we-pay-a-carbon-tax-to-our-own-government-or-to-someone-else-s-20210615-p5819q.html; "China Says EU's Planned Carbon Border Tax Violates Trade Principles," Reuters, July 26, 2021, https://www.reuters.com/business/sustainable-business/china-says-ecs-carbon-border-tax-is-expanding-climate-issues-trade-2021-07-26/; Adam Easton, "Turow: Vast Polish Coal Mine Infuriates the Neighbours," BBC News, June 17, 2021, https://www.bbc.com/news/world-europe-57484009.
23. Ben Rosario, "House Declares Climate Emergency: PH May Now Pursue Int'l Justice," *Manila Bulletin*, November 26, 2020, https://mb.com.ph/2020/11/26/house-declares-climate-emergency-ph-may-now-pursue-intl-justice/.
24. "China Warns Two-Thirds of Regions for Missing Energy Targets," Reuters, June 3, 2021, https://www.reuters.com/business/energy/china-warns-two-thirds-regions-missing-energy-targets-2021-06-03/; Michael McGowan, "Victorian Government Pledges to Slash State's Carbon Emissions by 50% by 2030," *Guardian*, May 2, 2021, https://www.theguardian.com/australia-news/2021/may/02/victorian-government-pledges-to-slash-states-carbon-emissions-by-50-by-2030; Sara Sneath, "Bill Seeks to Make Louisiana 'Fossil Fuel Sanctuary' in Bid against Biden's Climate Plans," *Guardian*, May 9, 2021, https://www.theguardian.com/us-news/2021/may/09/louisiana-bill-fossil-fuel-sanctuary; Yuka Obayashi and Sonali Paul, "Asia Snubs IEA's Call to Stop New Fossil Fuel Investments," Reuters, May 19, 2021, https://www.reuters.com/business/energy/asia-snubs-ieas-call-stop-new-fossil-fuel-investments-2021-05-19/.
25. "Poland Prolongs Turow Mine Life Despite International Outcry," Reuters, April 29, 2021, https://www.reuters.com/article/poland-coal-turow/poland-prolongs-turow-mine-life-despite-international-outcry-idUSL8N2MM6E9; "China Creates New 'Leaders Group' to Help Deliver Its Climate Goals," Carbon Brief, June 3, 2021, https://www.carbonbrief.org/explainer-china-creates-new

-leaders-group-to-help-deliver-its-climate-goals; Daniel O'Donoghue, "Ministers Grilled over Shetland Oil Project ahead of Climate Conference," Energy Voice, June 24, 2021, https://www.energyvoice.com/oilandgas/north-sea/332131/north-sea-cop26/; Branko Milanovic, "Is Norway the New East India Company?," *globalinequality* (blog), July 23, 2021, http://glineq.blogspot.com/2021/07/is-norway-new-east-india-company.html?m=1.

26. Charlotte Grieve, "'Big Mountain to Climb': Macquarie Chief Says Australia behind on EV Transition," *Sydney Morning Herald*, May 7, 2021, https://www.smh.com.au/business/banking-and-finance/macquarie-to-exit-coal-by-2024-hikes-dividend-20210507-p57pqb.html; Matthew Burgess, "A $213 Billion Investor Targets Whole Nation over Climate Change," Bloomberg, June 16, 2021, https://www.bloomberg.com/news/articles/2021-06-16/a-213-billion-investor-targets-whole-nation-over-climate-change; Sudarshan Varadhan and Nidhi Verma, "Indian Billionaires Face Off in Race to Solar Domination," Reuters, July 7, 2021, https://www.reuters.com/world/india/indian-billionaires-face-off-race-solar-domination-2021-07-07/.
27. Mission Innovation, http://mission-innovation.net/.
28. "China Internet Firms Must Commit to Clean Power as Consumption Soars—Greenpeace," Reuters, May 27, 2021, https://www.reuters.com/article/china-power-digital/china-internet-firms-must-commit-to-clean-power-as-consumption-soars-greenpeace-idUSL2N2NF01G; "2021 Progress Report to Parliament," Climate Change Committee, 2021, https://www.theccc.org.uk/publication/2021-progress-report-to-parliament/.
29. Ben Elgin, "A Top U.S. Seller of Carbon Offsets Starts Investigating Its Own Projects," Bloomberg, April 5, 2021, https://www.bloomberg.com/news/features/2021-04-05/a-top-u-s-seller-of-carbon-offsets-starts-investigating-its-own-projects; Tim Barsoe, "End of Wind Power Waste? Vestas Unveils Blade Recycling Technology," May 17, 2021, https://www.reuters.com/article/vestas-wind-technology-idCNL3N2N308L.
30. Bryan Alexander, *Academia Next: The Futures of Higher Education* (Baltimore: Johns Hopkins University Press, 2020), chapter 14; Paul Morland, *The Human Tide: How Population Shaped the Modern World* (New York: Public Affairs, 2019); Darrell Bricker, *Empty Planet: The Shock of Global Population Decline* (London: Robinson, 2019); Alexander, *Academia Next*, chapter 3.
31. Naomi Klein, *This Changes Everything: Capitalism vs. the Climate* (New York: Simon and Schuster, 2015), 443; Donna Haraway, *Staying with the Trouble: Making Kin in the Chthulucene* (Durham, NC: Duke University Press, 2016).
32. "Stop the Clock: The Environmental Benefits of a Shorter Working Week," Platform London, May 2021, https://6a142ff6-85bd-4a7b-bb3b-476b07b8f08d.usrfiles.com/ugd/6a142f_5061c06b240e4776bf31dfac2543746b.pdf.
33. M. Héry and M. Malenfer, "Development of a Circular Economy and Evolution of Working Conditions and Occupational Risks: A Strategic Foresight Study," *European Journal of Futures Research* 8, no. 8 (2020), https://doi.org/10.1186/s40309-020-00168-7.

34. George Monbiot, "'Green Growth' Doesn't Exist—Less of Everything Is the Only Way to Avert Catastrophe," *Guardian*, September 29, 2021, https://www.theguardian.com/commentisfree/2021/sep/29/green-growth-economic-activity-environment; Barbara Muraca, "Décroissance: A Project for a Radical Transformation of Society," in "Degrowth," special issue, *Environmental Values* 22, no. 2 (April 2013): 147–69, https://www.jstor.org/stable/23460976.
35. Shoshana Zuboff, *The Age of Surveillance Capitalism: The Fight for a Human Future at the New Frontier of Power* (New York: Profile Books, 2019); Klein, *This Changes Everything*, 255, 286–87.
36. Sam Meredith, "Carbon Capture Is Expected to Play a Pivotal Role in the Race to Net Zero Emissions. But Not Everyone Agrees," CNBC, July 20, 2021, updated July 23, 2021, https://www.cnbc.com/2021/07/20/climate-crisis-and-carbon-capture-why-some-are-worried-about-its-role.html; Matt Kessler, "The Environmental Cost of Internet Porn," *Atlantic*, December 13, 2017, https://www.theatlantic.com/technology/archive/2017/12/the-environmental-cost-of-internet-porn/548210/.
37. Geert Jan van Oldenborgh, Karin van der Wiel, Sarah Kew, Sjoukje Philip, Friederike Otto, Robert Vautard, Andrew King, et al., "Pathways and Pitfalls in Extreme Event Attribution," *Climatic Change* 166, no. 13 (2021), https://doi.org/10.1007/s10584-021-03071-7; "World Weather Attribution—Exploring the Contribution of Climate Change to Extreme Weather Events," https://www.worldweatherattribution.org/; Andreas Malm, *How to Blow Up a Pipeline: Learning to Fight in a World on Fire* (New York: Verso, 2021); Andreas Malm, *Corona, Climate, Chronic Emergency: War Communism in the Twenty-First Century* (New York: Verso, 2020); "SNP Activist Takes On UK Government in High Court over Oil and Gas," *The National* (Glasgow, Scotland), May 12, 2021, https://www.thenational.scot/news/19295641.snp-activist-takes-uk-government-high-court-oil-gas/.
38. Amanda Gearing, "How to Monitor the Bushfires Raging across Australia," The Conversation, January 3, 2020, https://theconversation.com/how-to-monitor-the-bushfires-raging-across-australia-129298; H. Tuba Özkan-Haller, "To Help Address the Climate Problem, Universities Must Rethink the Tenure and Promotion System," Yale Climate Connections, May 28, 2021, https://yaleclimateconnections.org/2021/05/to-help-address-the-climate-problem-universities-must-rethink-the-tenure-and-promotion-system/.
39. Chris Rapley, Sarah Bracking, Bill McGuire, Simon Lewis, and Jonathan Bamber, "Support a Science Oath for the Climate," *Guardian*, November 7, 2020, https://www.theguardian.com/science/2020/nov/07/support-a-science-oath-for-the-climate.
40. Michael Scott, "Free Speech at the Crossroads: International Dialogues—Who Will Speak for the Refugees?," webinar, Georgetown University, May 20, 2021; "Climate Emergency Declarations and Higher Education," Globally Responsible Leadership Initiative, updated September 16, 2019, https://responsibility.global/this-article-first-appeared-in-the-july-2019-grli-newsletter-and-will-be-updated-here-753e95780ed2.

Chapter 6. Best Case and Worst Case

Epigraphs. Jyoti Madhusoodanan, "Top US Scientist on Melting Glaciers: 'I've Gone from Being an Ecologist to a Coroner,'" *Guardian*, July 21, 2021, https://www.theguardian.com/environment/2021/jul/21/climate-crisis-glacier-diana-six-ecologist?CMP=oth_b-aplnews_d-1. Kim Stanley Robinson, *The Ministry for the Future* (New York: Orbit), 2020, 475.

1. Mark Z. Jacobson, Mark A. Delucchi, Zack A. F. Bauer, Savannah C. Goodman, William E. Chapman, Mary A. Cameron, Cedric Bozonnat, et al., "100% Clean and Renewable Wind, Water, and Sunlight All-Sector Energy Roadmaps for 139 Countries of the World," *Joule* 1, no. 1 (2017): 108–21, https://doi.org/10.1016/j.joule.2017.07.005.
2. Michael Grubb, Paul Drummond, Alexandra Poncia, Will McDowall, David Popp, Sascha Samadi, Cristina Penasco, et al., "Induced Innovation in Energy Technologies and Systems: A Review of Evidence and Potential Implications for CO_2 Mitigation," *Environmental Research Letters* 16, no. 4 (March 29, 2021), https://iopscience.iop.org/article/10.1088/1748-9326/abde07.
3. For one potential example, Ayar Al-zubaidi, Kenta Kobayashi, Yosuke Ishii, and Shinji Kawasaki, "One-Step Synthesis of Visible Light CO_2 Reduction Photocatalyst from Carbon Nanotubes Encapsulating Iodine Molecules," *Scientific Reports* 11 (2021): 10140, https://doi.org/10.1038/s41598-021-89706-2.
4. Sandeep Pai, Johannes Emmerling, Laurent Drouet, Hisham Zerriffi, and Jessica Jewell, "Meeting Well-Below 2°C Target Would Increase Energy Sector Jobs Globally," *One Earth* 4, no. 7 (2021): 1026–36, https://doi.org/10.1016/j.oneear.2021.06.005; Rupert Way, Matthew Ives, Penny Mealy, and J. Doyne Farmer, "Empirically Grounded Technology Forecasts and the Energy Transition," INET Oxford Working Paper No. 2021-01, September 14, 2021, https://www.inet.ox.ac.uk/files/energy_transition_paper-INET-working-paper.pdf.
5. Stefan Farsang, Marion Louvel, Chaoshuai Zhao, Mohamed Mezouar, Angelika D. Rosa, Remo N. Widmer, Xiaolei Feng, Jin Liu, and Simon A. T. Redfern, "Deep Carbon Cycle Constrained by Carbonate Solubility," *Nature Communications* 12 (2021), https://doi.org/10.1038/s41467-021-24533-7.
6. IPCC, *Climate Change 2021: The Physical Science Basis Summary for Policymakers* (Cambridge: Cambridge University Press, 2021), chttps://web.archive.org/web/20210902092728/https://www.ipcc.ch/report/ar6/wg1/downloads/report/IPCC_AR6_WGI_SPM.pdf.
7. Dan Lubin, Carl Melis, and David Tytler, "Ultraviolet Flux Decrease under a Grand Minimum from IUE Short-Wavelength Observation of Solar Analogs," *Astrophysical Journal Letters* 852, no. 1 (2018), https://iopscience.iop.org/article/10.3847/2041-8213/aaa124.
8. IPCC, *Climate Change 2021*; Lukoye Atwoli, Abdullah H Baqui, Thomas Benfield, Raffaella Bosurgi, Fiona Godlee, Stephen Hancocks, Richard Horton, et al., "Call for Emergency Action to Limit Global Temperature Increases, Restore Biodiversity, and Protect Health," *BMJ* 374 (2021), https://doi.org/10.1136/bmj.n1734.

9. IPCC, *Climate Change 2021*.
10. Christiana Figueres and Tom Rivett-Carnac, *The Future We Choose: Surviving the Climate Crisis* (New York: Vintage, 2020); Mark Maslin, "Climate Change: How Bad Could the Future Be If We Do Nothing?," The Conversation, May 6, 2021, https://theconversation.com/climate-change-how-bad-could-the-future-be-if-we-do-nothing-159665.
11. Kaj-Ivar van der Wijst, Andries F. Hof, and Detlef P. van Vuuren, "Costs of Avoiding Net Negative Emissions under a Carbon Budget," *Environmental Research Letters* 16, no. 6 (June 10, 2021), https://iopscience.iop.org/article/10.1088/1748-9326/ac03d9; Jem Bendell, "Deep Adaptation: A Map for Navigating Climate Tragedy," IFLAS Occasional Paper 2, July 27, 2018, revised July 27, 2020, https://lifeworth.com/deepadaptation.pdf.
12. David Roberts, "The Scariest Thing about Global Warming (and Covid-19)," Vox, December 4, 2020, https://www.vox.com/energy-and-environment/2020/7/7/21311027/covid-19-climate-change-global-warming-shifting-baselines.
13. L. Caesar, G. D. McCarthy, D. J. R. Thornalley, N. Cahill, and S. Rahmstorf, "Current Atlantic Meridional Overturning Circulation Weakest in Last Millennium," *Nature Geoscience* 14 (March 2021): 118–20, https://doi.org/10.1038/s41561-021-00699-z; Paul D. L. Ritchie, Greg S. Smith, Katrina J. Davis, Carlo Fezzi, Solmaria Halleck-Vega, Anna B. Harper, Chris A. Boulton, et al., "Shifts in National Land Use and Food Production in Great Britain after a Climate Tipping Point," *Nature Food* 1 (2020): 76–83, https://doi.org/10.1038/s43016-019-0011-3.
14. We can already identify possible early signs several of these enormously dangerous tipping points approaching: Luciana V. Gatti, Luana S. Basso, John B. Miller, Manuel Gloor, Lucas Gatti Domingues, Henrique L. G. Cassol, Graciela Tejada, et al., "Amazonia as a Carbon Source Linked to Deforestation and Climate Change," *Nature* 595 (2021): 388–93, https://doi.org/10.1038/s41586-021-03629-6; Niklas Boers and Martin Rypdal, "Critical Slowing Down Suggests That the Western Greenland Ice Sheet Is Close to a Tipping Point," *PNAS* 118, no. 21 (May 17, 2021): e2024192118, https://doi.org/10.1073/pnas.2024192118; A. K. Wåhlin, A. G. C. Grahamk, A. Hogan, Y. Queste, L. Boehme, R. D. Larter, E. C. Pettit, J. Wellner, and K. J. Heywood, "Pathways and Modification of Warm Water Flowing beneath Thwaites Ice Shelf, West Antarctica," *Science Advances* 7, no. 15 (April 9, 2021): https://doi.org/10.1126/sciadv.abd7254; Ian Joughin, Daniel Shapero, Ben Smith, Pierre Dutrieux, and Mark Barham, "Ice-Shelf Retreat Drives Recent Pine Island Glacier Speedup," *Science Advances* 7, no. 24 (June 11, 2021), https://doi.org/10.1126/sciadv.abg3080.
15. Kristofer Covey, Fiona Soper, Sunitha Pangala, Angelo Bernardino, and Zoe Pagliaro, "Carbon and Beyond: The Biogeochemistry of Climate in a Rapidly Changing Amazon," *Frontiers in Forests and Global Change* 4 (March 11, 2021), https://doi.org/10.3389/ffgc.2021.618401; Luciana V. Gatti, Luana S. Basso, John B. Miller, and Manuel Gloor, "Amazonia as a Carbon Source Linked to Deforestation and Climate Change," *Nature*, July 14, 2021, https://www.nature.com/articles/s41586-021-03629-6.epdf; Nico Wunderling, Jonathan F. Donges, Jürgen Kurths, and Ricarda Winkelmann, "Interacting Tipping Elements Increase Risk of Climate

Domino Effects under Global Warming," *Earth System Dynamics* 12 (2021): 601–19, https://doi.org/10.5194/esd-12-601-2021; Yang Chen, David M. Romps, Jacob T. Seeley, Sander Veraverbeke, William J. Riley, Zelalem A. Mekonnen, and James T. Randerson, "Future Increases in Arctic Lightning and Fire Risk for Permafrost Carbon," *Nature Climate Change* 11 (May 2021): 404–10, https://www.nature.com/articles/s41558-021-01011-y.epdf; Rebecca C. Scholten, Randi Jandt, Eric A. Miller, Brendan M. Rogers, and Sander Veraverbeke, "Overwintering Fires in Boreal Forests," *Nature* 593 (2021): 399–404, https://doi.org/10.1038/s41586-021-03437-y.

16. Elena Suglia, "Vanishing Nutrients," *Scientific American*, December 10, 2018, https://blogs.scientificamerican.com/observations/vanishing-nutrients/; Tamma A. Carleton, Amir Jina, Michael T. Delgado, Michael Greenstone, Trevor Houser, and Solomon M. Hsiang, "Valuing the Global Mortality Consequences of Climate Change Accounting for Adaptation Costs and Benefits," working paper 27599, National Bureau of Economic Research, July 2020, http://www.nber.org/papers/w27599.
17. "Dr. Rupert Read on Climate Catastrophe | UEA, UK | January 2019 | Extinction Rebellion UK" Extinction Rebellion UK, YouTube video, January 2019, https://youtu.be/rbzhc1BlvvI.
18. "Are "Net-Zero" Emissions a Smoke Screen?," The Analysis, November 30, 2020, https://theanalysis.news/are-net-zero-emissions-a-smoke-screen/.
19. Bill McKibben, *Falter: Has the Human Game Begun to Play Itself Out?* (New York: Holt McDougal, 2020); Naomi Klein, *This Changes Everything: Capitalism vs. the Climate* (New York: Simon and Schuster, 2015), 255, 286–87; Donna Haraway, *Staying with the Trouble: Making Kin in the Chthulucene* (Durham, NC: Duke University Press, 2016).

Chapter 7. What Is to Be Done

Epigraphs. David Wallace-Wells, "How to Live in a Climate 'Permanent Emergency,'" *New York Magazine*, July 1, 2021, https://nymag.com/intelligencer/2021/07/how-to-live-in-a-climate-permanent-emergency.html. IPCC, *Climate Change 2021: The Physical Science Basis Summary for Policymakers* (Cambridge: Cambridge University Press, 2021), chttps://web.archive.org/web/20210902092728/https://www.ipcc.ch/report/ar6/wg1/downloads/report/IPCC_AR6_WGI_SPM.pdf. Michael Mann and Tom Toles, *The Madhouse Effect* (New York: Columbia University Press, 2018), 13.

1. Jonathan Franzen on Facing the Facts of the Climate Crisis," Intelligence Squared, https://intelligencesquared.com/events/jonathan-franzen-on-facing-the-facts-of-the-climate-crisis-online/.
2. Banner seen on Red House grounds by author.
3. Other examples of campus planning include Walter Simpson, *Cool Campus! A How-To Guide for College and University Climate Action Planning* (Lexington, KY: Association for the Advancement of Sustainability in Higher Education, 2009), https://www.radford.edu/content/dam/departments/administrative/Sustainability/Documents/Cool-Campus-Climate-Planning-Guide.pdf; "Sustainable Campus," Cornell University, https://sustainablecampus.cornell.edu/; "Stanford Energy

System Innovations," Sustainable Stanford, https://sustainable.stanford.edu/campus-action/stanford-energy-system-innovations-sesi; "Sustainability," UC Davis, https://www.ucdavis.edu/about/sustainability/.

4. Elaine Newbern, "Rising to the Challenge: Eckerd's Plan to Combat Rising Sea Levels," *Current*, February 27, 2020, http://www.theonlinecurrent.com/news/rising-to-the-challenge-eckerd-s-plan-to-combat-rising-sea-levels/article_c7691d4c-5781-11ea-a65b-a7d50f6ad3cc.html/; "Eckerd 2100," Eckerd College, https://www.eckerd.edu/green/eckerd2100/; "UC San Diego Climate Action Plan," February 2019, https://sustain.ucsd.edu/_files/focus/UCSD-Climate-Action-Plan-2019-final.pdf.
5. Thanks to Joshua Kim and Chris Mayer for the thought during a blog post discussion: https://bryanalexander.org/book-club/our-ministry-for-the-future-reading-part-1/#comment-167530.
6. "Summary of Updates to Race to Zero criteria," Race to Zero, https://racetozero.unfccc.int/wp-content/uploads/2021/04/Summary-of-updates-to-Race-to-Zero-criteria.pdf.
7. "LOCKSS," Stanford University, https://www.lockss.org/.
8. Emily M. Bender, Timnit Gebru, Angelina McMillan-Major, and Shmargaret Shmitchell, "On the Dangers of Stochastic Parrots: Can Language Models Be Too Big?," *FAccT '21: Proceedings of the 2021 ACM Conference on Fairness, Accountability, and Transparency* (March 2021): 610–23, https://doi.org/10.1145/3442188.3445922; Clive Thompson, "A Gloriously Fixable Laptop," Debugger, Medium, August 27, 2021, https://debugger.medium.com/a-gloriously-fixable-laptop-e87a2cbd76ea; Roy Schwartz, Jesse Dodge, Noah A. Smith, and Oren Etzioni, "Green AI," arXiv:1907.10597v3 [cs.CY], August 13, 2019, https://arxiv.org/pdf/1907.10597.pdf; Peter Henderson, Jieru Hu, Joshua Romoff, Emma Brunskill, Dan Jurafsky, Joelle Pineau, "Towards the Systematic Reporting of the Energy and Carbon Footprints of Machine Learning," *Journal of Machine Learning Research* 21 (2020): 1–43, https://jmlr.org/papers/volume21/20-312/20-312.pdf; Nico Wunderling, Johnathan F. Donges, Jürgen Kurths, and Ricarda Winkelmann, "Interacting Tipping Elements Increase Risk of Climate Domino Effects under Global Warming," *Earth System Dynamics* 12 (2021): 601–19, https://doi.org/10.5194/esd-12-601-2021.
9. Rebecca Schmid, "How Operas Are Going Green," *New York Times*, May 10, 2021, https://www.nytimes.com/2021/05/10/arts/music/opera-sustainability.html; Paul Hawken, *Drawdown: The Most Comprehensive Plan Ever Proposed to Reverse Global Warming* (New York: Penguin, 2017), 96; David Bornstein, "Investing in Energy Efficiency Pays Off," *New York Times*, February 6, 2015, https://opinionator.blogs.nytimes.com/2015/02/06/investing-in-energy-efficiency-pays-off/.
10. Lee Gardner, "For Colleges, Climate Change Means Making Tough Choices," *Chronicle of Higher Education*, May 3, 2019, https://www.chronicle.com/article/for-colleges-climate-change-means-making-tough-choices/.
11. Ayodele Johnson, "How Africa's Largest City Is Staying Afloat," BBC Future Planet, January 21, 2021, https://www.bbc.com/future/article/20210121-lagos-nigeria-how-africas-largest-city-is-staying-afloat.
12. "Living Building Challenge," International Living Future Institute, https://living-future.org/lbc/; Hawken, *Drawdown*, 188–89.

13. Lidia Morawska, Joseph Allen, William Bahnfleth, Philomena M. Bluyssen, Atze Boerstra, Giorgio Buonanno, Junji Cao, et al., "A Paradigm Shift to Combat Indoor Respiratory Infection," *Science* 372, no. 6543 (May 2021), https://www.science.org/doi/10.1126/science.abg2025; "Schemmel Theology Lecture—Dan Dileo," Clarke A/V, YouTube video, March 23, 2021, https://youtu.be/z261QoUuoaM; Florian Ludeke-Freund and Simon Burandt, "Universities as Learning Organizations for Sustainability? The Task of Climate Protection," in Leal Filho, ed., *Universities and Climate Change: Introducing Climate Change to University Programmes* (Heidelberg, Germany: Springer, 2010), 188; Lilah Burke, "Preparing for a Crisis," Inside Higher Ed, May 5, 2021, https://www.insidehighered.com/news/2021/05/05/colleges-brace-effects-climate-change-campus.
14. "Work Underway on Substantial Solar Farm near Campus," Grinnell College, August 19, 2021, https://alumni.grinnell.edu/news/solar-farm; Nav Sangha, "Red Deer College Now Has 4,200 Solar Panels across Campus," CTV News, June 8, 2021, updated June 9, 2021; https://edmonton.ctvnews.ca/red-deer/red-deer-college-now-has-4-200-solar-panels-across-campus-1.5461980; Claire Potter, "Royalton-Area Solar Project Aims for Affordability," *Valley News* (West Lebanon, NH), July 18, 2021, https://www.vnews.com/Construction-starts-at-Royalton-community-solar-project-41450042; Maria Gallucci, "A Different Kind of Solar Technology Is Poised to Go Big," Grist, July 26, 2021, https://grist.org/energy/cadmium-telluride-technology-first-solar/.
15. "Berea College Brings Hydroelectric Power to Region," *Lane Report*, September 22, 2021, https://www.lanereport.com/146689/2021/09/berea-college-brings-hydroelectric-power-to-region/; "Largest Anaerobic Digester in the Northeast Begins Production of Renewable Energy," Middlebury College, July 21, 2021, https://www.middlebury.edu/newsroom/archive/2021-news/node/658771; Hawken, *Drawdown*, 15; "Bioenergy Research Demonstration Facility (BRDF)," University of British Columbia Energy and Water Services, https://energy.ubc.ca/ubcs-utility-infrastructure/brdf/; Chris Riedy and Jane Daly, "Targeting a Low-Carbon University: A Greenhouse Gas Reduction Target for the Australian Technology Network of Universities," in Leal Filho, *Universities and Climate Change*, 154–55.
16. Marcus Fairs, "Planting Trees "Doesn't Make Any Sense" in the Fight against Climate Change Due to Permanence Concerns, Say Experts," Dezeen, July 5, 2021, https://www.dezeen.com/2021/07/05/carbon-climate-change-trees-afforestation/; Javier Beneyas, Inmaculada Alonso, David Alba, and Luis Pertierra, "The Impact of Universities on the Climate Change Process," in Leal Filho, *Universities and Climate Change*, 51–54. "Mangroves Are Trees and Shrubs That Have Adapted to Life in a Saltwater Environment," Florida Keys National Marine Sanctuary, https://floridakeys.noaa.gov/plants/mangroves.html; Elizabeth Kolbert, *Under a White Sky: The Nature of the Future* (New York: Crown, 2021), 161; "Mental Health and Our Changing Climate: Impacts, Implications, and Guidance," American Psychological Association, March 2017, https://www.apa.org/news/press/releases/2017/03/mental-health-climate.pdf. One leading exemplar of biophilic design is Terrapin, and one example of their work is "The Nature Of Wood: An Exploration of the Science on Biophilic Responses to Wood," Terrapin, 2022, http://www

.terrapinbrightgreen.com/wp-content/uploads/2022/01/The-Nature-of-Wood_Terrapin_2022-01.pdf.

17. Anand Kulkarni, "In an Era of Disruption, Do We Need a Resilience Ranking?," University World News, December 12, 2020, https://www.universityworldnews.com/post.php?story=20201208085241207.
18. Kolbert, *Under a White Sky*, 158–59; Hawken, *Drawdown*, 176–77; Susan Cosier, "How Adding Rock Dust to Soil Could Help Get Carbon into the Ground," Yale Environment 360, September 2, 2021, https://e360.yale.edu/features/how-adding-rock-dust-to-soil-can-help-get-carbon-into-the-ground.
19. Z. Ai, N. Hanasaki, and V. Heck, et al., "Global Bioenergy with Carbon Capture and Storage Potential Is Largely Constrained by Sustainable Irrigation," *Nature Sustainability* 4 (2021), 884–91, https://doi.org/10.1038/s41893-021-00740-4; Kolbert, *Under a White Sky*, 163–64.
20. "Global Methane Assessment: Benefits and Costs of Mitigating Methane Emissions," United Nations Environment Programme, 2021, https://www.ccacoalition.org/en/file/7941/download?token=q_bCnfYV.
21. Cathy Gere and Adam Aron, "Carbon Neutrality at the University of California Is a Form of Climate Change Denial," Remaking the Campus, November 26, 2020, https://utotherescue.blogspot.com/2020/11/carbon-neutrality-at-university-of.html; Naomi Klein, *This Changes Everything: Capitalism vs. the Climate* (New York: Simon and Schuster, 2015), 354, 401; Daisy Dunne and Robbie Mallett, "Russell Group Universities Received £60M in Funding from Coal, Oil and Gas Sector in Last Five Years," *Independent*, November 29, 2020, https://www.independent.co.uk/climate-change/news/russell-group-universities-received-ps60m-in-funding-from-coal-oil-and-gas-sector-in-last-five-years-b1761160.html; Jessica Colarossi, "Boston University to Divest from Fossil Fuel Industry," BU Today, September 23, 2021, https://www.bu.edu/articles/2021/boston-university-divest-from-fossil-fuel-industry/.
22. "Redesigning a University," The Future Trends Forum, YouTube video, Feb 25, 2021, https://youtu.be/Xj_rqIEI5MA; Solitaire Townsend, "We Urgently Need 'Scope X' Business Leadership for Climate," *Forbes*, June 29, 2020, https://www.forbes.com/sites/solitairetownsend/2020/06/29/we-urgently-need-scope-x-business-leadership-for-climate/?sh=48babbcb4dd3.
23. "An Electrifying Partnership," Sourcewell, January28, 2021, https://news.sourcewell-mn.gov/an-electrifying-partnership?eml=20210809; Tom Vander Ark, "The Electric Future of Education," *Forbes*, June 7, 2021, https://www.forbes.com/sites/tomvanderark/2021/06/07/the-electric-future-of-education-transportation/?sh=249cf1f13bab; Javier Beneyas, Inmaculada Alonso, David Alba, and Luis Pertierra, "The Impact of Universities on the Climate Change Process," in Leal Filho, *Universities and Climate Change*, 58; Millie Rooney and Jennifer McMillin, "The Campus as a Classroom: Integrating People, Place, and Performance for Communicating Climate Change" in Leal Filho, *Universities and Climate Change*, 127.
24. Tim Hepher and Laurence Frost, "Airbus Tells EU Hydrogen Won't Be Widely Used in Planes before 2050," Reuters, June 10, 2021, https://www.reuters.com

/business/aerospace-defense/airbus-tells-eu-hydrogen-wont-be-widely-used -planes-before-2050-2021-06-10/; Hans de Wit and Philip G Altbach, "Time to Cut International Education's Carbon Footprint," University World News, January 11, 2020, https://www.universityworldnews.com/post.php?story=20200108084344396; Jenny J. Lee and Ola A. Lundemo, "Why Sustainability Is Not Enough in International Education," University World News, June 5, 2021, https://www.universityworldnews.com/post.php?story=20210531123436636; Wesley Jenkins, "The Thanksgiving–Winter Break Boomerang: Is All That Travel Really Necessary?," *Chronicle of Higher Education*, December 10, 2019, https://www.chronicle.com/article/the-thanksgiving-winter-break-boomerang-is-all-that-travel-really-necessary/.

25. Grant Faber, "A Framework to Estimate Emissions from Virtual Conferences," *International Journal of Environmental Studies* 78, no. 4 (2021): 608–23, https://doi.org/10.1080/00207233.2020.1864190. See this Twitter conversation between Jeroen Bosman and Tanja de Bie, starting here: https://twitter.com/jeroenbosman/status/1215747230713278464.

26. Inside Higher Ed, "Pilgrimages Could Save the Environment," *Academic Minute*, June 11, 2020, https://www.insidehighered.com/audio/2020/06/11/pilgrimages-could-save-environment; Ellie Bothwell, "British Universities Urged to Limit Flights," Times Higher Education/Inside Higher Education, November 27, 2019, https://www.insidehighered.com/news/2019/11/27/british-universities-urged-limit-flights-reduce-carbon-imprint.

27. Volker Grewe, Arvind Gangoli Rao, Tomas Grönstedt, Carlos Xisto, Florian Linke, Joris Melkert, Jan Middel, et al., "Evaluating the Climate Impact of Aviation Emission Scenarios towards the Paris Agreement Including COVID-19 Effects," *Nature Communications* 12 (2021): 3841, https://doi.org/10.1038/s41467-021-24091-y.

28. Richard B. Alley, *The Two-Mile Time Machine: Ice Cores, Abrupt Climate Change, and Our Future* (Princeton, NJ: Princeton University Press, 2014); Brady Dennis, "In Fast-Warming Minnesota, Scientists Are Trying to Plant the Forests of the Future," *Washington Post*, April 29, 2020, https://www.washingtonpost.com/graphics/2020/climate-solutions/climate-change-minnesota/.

29. Xiangyu Li, Joseph Peoples, Peiyan Yao, and Xiulin Ruan, "Ultrawhite BaSO4 Paints and Films for Remarkable Daytime Subambient Radiative Cooling," *Applied Materials and Interfaces* 13, no. 18 (2021): 21733–39, https://doi.org/10.1021/acsami.1c02368; Brian Fagan, *The Attacking Ocean: The Past, Present, and Future of Rising Sea Levels* (New York: Bloomsbury, 2013), 236; Adele Peters, "Could Pulling Water from the Air Be the Solution for California's Drought-Stricken Towns?," *Fast Company*, August 13, 2021, https://www.fastcompany.com/90665075/could-pulling-water-from-the-air-be-the-solution-for-californias-drought-stricken-towns; Laura Paddison, "How Replanted Seagrass Is Restoring the Ocean," *Yes! Magazine*, Aug 2, 2021, https://www.yesmagazine.org/environment/2021/08/02/replanted-seagrass-is-restoring-the-ocean.

30. Christopher J. Ryan and Christopher R. Marsicano, "Examining the Impact of Divestment from Fossil Fuels on University Endowments," *New York University*

Journal of Law and Business 17 (2020): 95–152, https://papers.ssrn.com/sol3/papers.cfm?abstract_id=3501231#; Bernard Feltz, "The Philosophical and Ethical Issues of Climate Change," *UNESCO Courier* (2019), https://en.unesco.org/courier/2019-3/philosophical-and-ethical-issues-climate-change; Niina Kautto, Alexei Trundle, and Darryn McEvoy, "Climate Adaptation Planning in the Higher Education Sector," *International Journal of Sustainability in Higher Education* 19, no. 7 (2018), https://www.emerald.com/insight/content/doi/10.1108/IJSHE-02-2018-0028/full/html; Leal Filho, *Universities and Climate Change*; Walter Leal Filho, ed., *Climate Change Research at Universities: Addressing the Mitigation and Adaptation Challenges* (Heidelberg: Springer, 2017).

31. Ma Haiyan, "Tsinghua University Established Carbon Neutral Research Institute," China News Network, September 22, 2021, https://m.chinanews.com/wap/detail/zw/gn/2021/09-22/9571286.shtml; "The Environmental Change Institute," University of Oxford, https://www.eci.ox.ac.uk/; "About," Johns Hopkins Ralph O'Connor Sustainable Energy Institute, https://energyinstitute.jhu.edu/about/; "GeoMIP Welcome," https://climate.envsci.rutgers.edu/GeoMIP/; "Centre for Climate Repair at Cambridge," University of Cambridge, https://www.climaterepair.eng.cam.ac.uk/; Youngsuk "YS" Chi (chairman of Elseview, director of corporate affairs and strategy for RELX), interview by the author, 2021.
32. Giacomo Grassi, Elke Stehfest, Joeri Rogelj, Detlef van Vuuren, Alessandro Cescatti, Jo House, Gert-Jan Nabuurs, et al., "Critical Adjustment of Land Mitigation Pathways for Assessing Countries' Climate Progress," *Nature Climate Change* 11 (2021), 425–34, https://doi.org/10.1038/s41558-021-01033-6; "NASA-Built Instrument Will Help to Spot Greenhouse Gas Super-Emitters," Jet Propulsion Lab, April 15, 2021, https://www.jpl.nasa.gov/news/nasa-built-instrument-will-help-to-spot-greenhouse-gas-super-emitters; "Who We Are," Carbon Mapper, https://carbonmapper.org/about-us/; "Renewable Energy," Muhlenberg College, https://www.muhlenberg.edu/offices/sustainability/programs/plantoperations/renewableenergy/.
33. Barsoe, "End of Wind Power Waste?"; "About," ClimatePrediction.net, https://www.climateprediction.net/about/; Youngsuk "YS" Chi interview, 2021.
34. Saijel Kishan and Eric Roston, "What Climate Science Loses without Enough Black Researchers," Bloomberg, May 27, 2021, https://www.bloomberg.com/news/features/2021-05-27/addressing-racism-inside-climate-science.
35. Peter Reeves, Sophie Mason, and Luke Sanders, "A National Study of Capital Infrastructure at Colleges and Schools of Agriculture: An Update Produced by Gordian," Association of Public and Land-Grant Universities, February 2021, https://www.aplu.org/library/a-national-study-of-capital-infrastructure-at-colleges-and-schools-of-agriculture-an-update/file.
36. "Framework for Information Literacy for Higher Education," Association of College and Research Libraries, 2015, https://www.ala.org/acrl/standards/ilframework; D. Rosowsky, "Climate and Change: Reflections on *The Sixth Extinction*," fall 2016, https://www.davidvrosowsky.com/wp-content/uploads/2020/07/DVR-SIXTH-EXTINCTION-ESSAY.pdf.

37. Susan Irais, "By 2040, 80% of Energy Used on Tec de Monterrey Campuses Will Be Renewable," Institute for the Future of Education Observatory, September 20, 2021, https://observatory.tec.mx/edu-news/tec-de-monterrey-2025-sustainability-and-climate-change-plan.
38. Nathanael Johnson, "Remember Snow Days? Today's Kids Get Heat Days," Grist, June 14, 2021, https://grist.org/extreme-weather/remember-snow-days-todays-kids-get-heat-days/.
39. Walter Leal Filho, "Climate Change at Universities: Results of a World Survey," in Leal Filho, *Universities and Climate Change*, 3.
40. "Oregon Health Authority and Confederated Tribes of Warm Springs: Using Storytelling to Illustrate the Impacts of Climate Change on Health," StoryCenter, https://www.storycenter.org/case-studies//oregon-health-authority-and-confederated-tribes-of-warm-springs-using-storytelling-to-illustrate-the-impacts-of-climate-change-on-health.
41. Phil Hill, "The Colleges That Prospered during the Pandemic: A Common Theme Unites Arizona State, Central Florida, Kennesaw State, and Others That Have Thrived," *Chronicle of Higher Education*, July 28, 2021, https://www.chronicle.com/article/the-colleges-that-prospered-during-the-pandemic; Ian Bogost, "America Will Sacrifice Anything for the College Experience," *Atlantic*, October 20, 2020, https://www.theatlantic.com/technology/archive/2020/10/college-was-never-about-education/616777/; Audrey Watters, "Hope for the Future," Hack Education, March 8, 2022, http://hackeducation.com/2022/03/08/hope.
42. Ashlee Cunsolo, Sherilee L. Harper, Kelton Minor, Katie Hayes, Kimberly G. Williams, and Courtney Howard, "Ecological Grief and Anxiety: The Start of a Healthy Response to Climate Change?," *Lancet* 4 (July 2020), https://www.thelancet.com/action/showPdf?pii=S2542-5196%2820%2930144-3; Namrata Chowdhary, "Acknowledging the Weight of the Current Moment: Grief and the Climate Crisis," 350.org, Medium, May 19, 2021, https://350.medium.com/acknowledging-the-weight-of-the-current-moment-grief-and-the-climate-crisis-5c035624c4b5; "Climate Cafes," https://climatecafes.org/; Leal Filho, "Climate Change at Universities," in Leal Filho, *Universities and Climate Change*, 7.
43. "What Is a Work College?," Work Colleges Consortium, https://workcolleges.org/what-work-college; Kevin Maher, "A Year Later, a Shroud of Uncertainty Still Surrounds UF's COVID-19 Numbers," *Independent Florida Alligator*, April 19, 2021, https://www.alligator.org/article/2021/04/a-year-later-a-shroud-of-uncertainty-still-surrounds-ufs-covid-19-numbers.
44. "The Domes," Sustainable Living and Learning Communities, https://sllc.ucdavis.edu/domes.
45. "The Peabody Essex Museum's Climate + Environment Initiative Seeks to Create a Different Climate Future," American Alliance of Museums, July 16, 2021, https://www.aam-us.org/2021/07/16/the-peabody-essex-museums-climate-environment-initiative-seeks-to-create-a-different-climate-future/.
46. Stephen Nash, *Virginia Climate Fever: How Global Warming Will Transform Our Cities, Shorelines, and Forests* (Charlottesville: University of Virginia Press, 2014),

64–65; Emily Atkin, "The Climate Case for Abolition," Heated, April 21, 2021, https://heated.world/p/the-climate-case-for-abolition?s=r.

47. Bryan Alexander, *Academia Next: The Futures of Higher Education* (Baltimore: Johns Hopkins University Press, 2020), 226–27; "International Universities Climate Alliance," https://www.universitiesforclimate.org/; Maruf Sanni, James O. Adejuwon, Idowu Ologh, and William O. Siyanbola, "Path to the Future for Climate Change Education: A University Project Approach," in Leal Filho, *Universities and Climate Change*, 21–30; Dagnija Blumberga and Maris Klavis, "Climate Change Education in the Curricula of Technical and Classical Universities," in Leal Filho, *Universities and Climate Change*, 100–105; Chris Riedy and Jane Daly, "Targeting a Low-Carbon University: A Greenhouse Gas Reduction Target for the Australian Technology Network of Universities," in Leal Filho, *Universities and Climate Change*, 151–62; "Campus. Community. Global," Second Nature, https://secondnature.org/; "The Cooperative Institute for Climate, Ocean, and Ecosystem Studies," https://cicoes.uw.edu/; "EcoLeague," https://ecoleague.org/.

48. "Sustainable Endowments Institute," https://www.endowmentinstitute.org/; "Adaptation Learning Network," https://adaptationlearningnetwork.com/about-us-1; "Collaborative Heat Watch Research Project," Virginia Foundation for Independent Colleges, https://www.vfic.org/vfic-programs/stem-faculty-initiative/collaborative-heat-mapping-research-project/; "VFIC Schools and VSU Heat Watch Taking Place on July 15," Virginia Foundation for Independent Colleges, July 7, 2021, https://drive.google.com/file/d/1gHK3L1MYCNOcoKCwcgNQoHtWK-LztY_U/view.

49. "Scientists' Statement on Lowering Atmospheric Methane Concentrations," Methane Action, April 16, 2021, https://methaneaction.org/expert-statement-oxidation-methane/; National Academies of Sciences, Engineering, and Medicine, *Reflecting Sunlight: Recommendations for Solar Geoengineering Research and Research Governance* (Washington, DC: National Academies Press, 2021), https://nap.nationalacademies.org/catalog/25762/reflecting-sunlight-recommendations-for-solar-geoengineering-research-and-research-governance; William J. Ripple, Christopher Wolf, Thomas M. Newsome, Phoebe Barnard, and William R. Woomaw, "World Scientists' Warning of a Climate Emergency," *BioScience* 70, no. 1 (2019), https://doi.org/10.1093/biosci/biz088.

50. "Universities Declare Climate Emergency ahead of Climate Action Summit," International Institute for Sustainable Development, July 23, 2019, https://sdg.iisd.org/news/universities-declare-climate-emergency-ahead-of-climate-action-summit/; "The 17 Goals," Department of Economic and Social Affairs Sustainable Development, https://sdgs.un.org/goals.

51. Jem Bendell, "Deep Adaptation: A Map for Navigating Climate Tragedy," IFLAS Occasional Paper 2, July 27, 2018, revised July 27, 2020, https://lifeworth.com/deepadaptation.pdf.

52. Faith E. Pinho and Alex Wigglesworth, "California's Drought and Wildfire Dangers Rising at Stunning Pace," *Los Angeles Times*, June 26, 2021, https://www.latimes.com/california/story/2021-06-26/drought-wildfire-conditions-evolving-at-unprecedented-pace.

53. Robinson Meyer, "How the U.S. Made Progress on Climate Change without Ever Passing a Bill," *Atlantic*, June 16, 2021, https://www.theatlantic.com/science/archive/2021/06/climate-change-green-vortex-america/619228/.
54. Feargus O'Sullivan, "Berlin's New Timber Tower Comes with Lofty Ambitions," Bloomberg, February 11, 2021, https://www.bloomberg.com/news/articles/2021-02-11/berlin-has-high-hopes-for-this-wooden-skyscraper.
55. Rapid Transition Alliance Staff, "Educating Girls Is More Effective in the Climate Emergency than Many Green Technologies," Resilience, May 26, 2020, https://www.resilience.org/stories/2020-05-26/educating-girls-is-more-effective-in-the-climate-emergency-than-many-green-technologies-2/; Hawken, *Drawdown.*
56. It is easy to find examples of youth climate activism, starting with Greta Thunberg. Gideon Polya, "Millions Join Global School Climate Strike—We Are Running Out of Time in Climate Change," Counter Currents, September 22, 2019, https://countercurrents.org/2019/09/millions-join-global-school-climate-strike-we-are-running-out-of-time/; Fridays for Future, https://fridaysforfuture.org/.
57. Jeff Bezos, "Bezos Earth Fund," Instagram, November 16, 2020, https://www.instagram.com/p/CHpwxnvHufz/; Regine Clement, "Catalytic Capital," in *All We Can Save: Truth, Courage, and Solutions for the Climate Crisis*, ed. Katharine K. Wilkinson and Ayana Elizabeth Johnson (New York: Penguin, 2020), 171–76.
58. L. T. Keyßer and M. Lenzen, "1.5°C Degrowth Scenarios Suggest the Need for New Mitigation Pathways," *Nature Communications* 12 (2021): 2676, https://doi.org/10.1038/s41467-021-22884-9; A. Frank, Jonathan Carroll-Nellenback, M. Alberti, and A. Kleidon, "The Anthropocene Generalized: Evolution of Exo-civilizations and Their Planetary Feedback," *Astrobiology* 18, no. 5 (May 2018), https://doi.org/10.1089/ast.2017.1671.
59. Viviane Clement, Kanta Kumari Rigaud, Alex de Sherbinin, Bryan Jones, Susana Adamo, Jacob Schewe, Nian Sadiq, and Elham Shabahat, *Groundswell, Part 2: Acting on Internal Climate Migration* (Washington, DC: World Bank, 2021), https://openknowledge.worldbank.org/handle/10986/36248.
60. "Letter from Coalition of Organizations to Secretary of State and HHS on University Sponsorship of Refugees," Presidents' Alliance on Higher Education and Immigration, March 9, 2021, https://www.presidentsalliance.org/letter-university-sponsorship/; Michael Scott, "Free Speech at the Crossroads: International Dialogues—Who Will Speak for the Refugees?," webinar, Georgetown University, May 20, 2021.
61. A. K. Streeter, "Detroit Will Be a Climate Refuge for City Dwellers . . . in 2100," Treehugger, September 23, 2014, https://web.archive.org/web/20140926155532/http:/www.treehugger.com/climate-change/detroit-will-be-climate-refuge-citydwellers-2100.html.
62. Peter Sutoris, "Anthropocene Skills Need to Be at the Centre of Curricula," University World News, January 30, 2021, https://www.universityworldnews.com/post.php?story=20210127065228110.
63. Trent Batson, "Universities Should Lead the Fight to Preserve Humanity," University World News, January 25, 2020, https://www.universityworldnews.com

/post.php?story=20200121142658618; Trent Batson, *The Last Humans Project: Humans Surviving the Climate Extinction Event*, Release 1.0, January 2020, https://ad897497-52dd-4917-9ba2-72ec2b0b323e.filesusr.com/ugd/6f8c7f_c747de52461d45789c4f89804471od5d.pdf.

64. Jainey K. Bavishi, "A Tale of Three Cities," in Wilkinson and Johnson, *All We Can Save*, 157–65; Kate Orff, "Mending the Landscape," in Wilkinson and Johnson, *All We Can Save*, 180.

INDEX

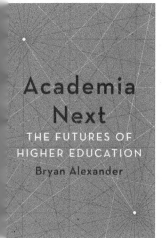

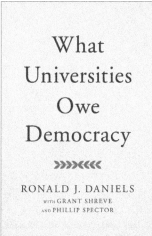

HIGHER EDUCATION BOOKS FROM HOPKINS PRESS

Academia Next
The Futures of Higher Education
Bryan Alexander
2020 Most Significant Futures Work Award Winner, Association of Professional Futurists

What Universities Owe Democracy
Ronald J. Daniels with Grant Shreve & Phillip Spector
"The methods to defend democracy must be taught, and What Universities Owe Democracy is our textbook."
—Garry Kasparov, Chairman of the Renew Democracy Initiative / former World Chess Champion

Alternative Universities
Speculative Design for Innovation in Higher Education
David J. Staley
"This book gives us the toolkit we need to open our minds to the possible university: the one that [illegible] us. We can must push ourselves to move beyond data-driven decision-making and let our imaginations loose in the present." —Maree Conway, Founder, Thinking Futures

press.jhu.edu

@JohnsHopkinsUniversityPress

@HopkinsPress

@JHUPress